T0205355

CMOS FRACTIONAL-N SYNTHESIZERS

THE KLUWER INTERNATIONAL SERIES IN ENGINEERING AND COMPUTER SCIENCE

ANALOG CIRCUITS AND SIGNAL PROCESSING
Consulting Editor: Mohammed Ismail. *Ohio State University*

Related Titles:

MODULAR LOW-POWER, HIGH SPEED CMOS ANALOG-TO-DIGITAL CONVERTER FOR EMBEDDED SYSTEMS
Lin, Kemna & Hosticka
ISBN: 1-4020-7380-1
DESIGN CRITERIA FOR LOW DISTORTION IN FEEDBACK OPAMP CIRCUITE
Hernes & Saether
ISBN: 1-4020-7356-9
CIRCUIT TECHNIQUES FOR LOW-VOLTAGE AND HIGH-SPEED A/D CONVERTERS
Walteri
ISBN: 1-4020-7244-9
DESIGN OF HIGH-PERFORMANCE CMOS VOLTAGE CONTROLLED OSCILLATORS
Dai and Harjani
ISBN: 1-4020-7238-4
CMOS CIRCUIT DESIGN FOR RF SENSORS
Gudnason and Bruun
ISBN: 1-4020-7127-2
ARCHITECTURES FOR RF FREQUENCY SYNTHESIZERS
Vaucher
ISBN: 1-4020-7120-5
THE PIEZOJUNCTION EFFECT IN SILICON INTEGRATED CIRCUITS AND SENSORS
Fruett and Meijer
ISBN: 1-4020-7053-5
CMOS CURRENT AMPLIFIERS; SPEED VERSUS NONLINEARITY
Koli and Halonen
ISBN: 1-4020-7045-4
MULTI-STANDARD CMOS WIRELESS RECEIVERS
Li and Ismail
ISBN: 1-4020-7032-2
A DESIGN AND SYNTHESIS ENVIRONMENT FOR ANALOG INTEGRATED CIRCUITS
Van der Plas, Gielen and Sansen
ISBN: 0-7923-7697-8
RF CMOS POWER AMPLIFIERS: THEORY, DESIGN AND IMPLEMENTATION
Hella and Ismail
ISBN: 0-7923-7628-5
DATA CONVERTERS FOR WIRELESS STANDARDS
C. Shi and M. Ismail
ISBN: 0-7923-7623-4
DIRECT CONVERSION RECEIVERS IN WIDE-BAND SYSTEMS
A.   Parssinen
ISBN: 0-7923-7607-2
AUTOMATIC CALIBRATION OF MODULATED FREQUENCY SYNTHESIZERS
D. McMahill
ISBN: 0-7923-7589-0
MODEL ENGINEERING IN MIXED-SIGNAL CIRCUIT DESIGN
S. Huss
ISBN: 0-7923-7598-X
ANALOG DESIGN FOR CMOS VLSI SYSTEMS
F. Maloberti
ISBN: 0-7923-7550-5
CONTINUOUS-TIME SIGMA-DELTA MODULATION FOR A/D CONVERSION IN RADIO RECEIVERS   L. Breems, J.H. Huijsing
ISBN: 0-7923-7492-4
DIRECT DIGITAL SYNTHESIZERS: THEORY, DESIGN AND APPLICATIONS
J. Vankka, K. Halonen
ISBN: 0-7923 7366-9

CMOS FRACTIONAL-N SYNTHESIZERS

Design for High Spectral Purity and Monolithic Integration

by

Bram De Muer

KU Leuven, Belgium

and

Michiel Steyaert

KU Leuven, Belgium

SPRINGER-SCIENCE+BUSINESS MEDIA, B.V.

A C.I.P. Catalogue record for this book is available from the Library of Congress.

ISBN 978-1-4419-5343-8   ISBN 978-0-306-48001-0 (eBook)
DOI 10.1007/978-0-306-48001-0

Printed on acid-free paper

Abstract

The years around the turn of the century will be remembered for the unrivaled growth of mobile telecommunications. Driven by synergetic economical and technological forces, mobile devices have become mass-consumer products whose size and cost is continually decreasing for increasing functionality. While the digital back-end of the mobile phones today is highly-integrated, i.e. small and cost-effective, the RF transceiver part still relies on expensive external components.

The presented research fits in the quest for small and cheap cellular transceiver solutions, which could ultimately lead to the single-chip mobile phone. Therefore, monolithic integration has been targeted to reduce the number of components (i.e. the cost) and that in the cheapest technology: CMOS. A major challenge in CMOS single-chip transceivers is the frequency synthesizer that generates the local oscillator signal. Its spectral purity (phase noise, spurious tones) is critical for the quality of the information transfer. Its dynamic behavior determines the maximal information throughput, which is the main reason of existence of 2.5G and 3G systems.

This book is conceived as a manual for the design of monolithic CMOS frequency synthesizers, focused on attaining the highest spectral purity and fast switching with moderate power consumption , with ultimately the successful implementation of a monolithic CMOS $\Delta\Sigma$-controlled fractional-N frequency synthesizer. $\Delta\Sigma$ fractional-N synthesis has been selected, since it trades off accuracy for digital complexity, i.e. the natural biotope of CMOS.

The first obstacles towards monolithic integration of the PLL are the high-speed prescaler and oscillator (VCO). Dual- and multi-modulus prescaler design is elaborated focusing on the search for the best high-speed flipflop over different specifications: speed (up to 15 GHz) and power, but also residual phase noise and input sensitivity. In VCO design, the phase noise is crucial; A design-oriented phase noise theory is developed that takes into account the non-linearity of the active elements. An in-house simulator-optimizer program is developed and applied to integrate high-Q inductors on-chip. 1/f noise upconversion mechanisms have been identified and successfully countered in VCO circuit design. A VCO bias filtering technique has been experimentally verified that removes all flicker noise.

A systematic design strategy has been developed that marks out the design space for the PLL loop filter, trading off all noise contributions and fast dynamics for integrated capacitance (area). A dual-path loop filter is developed that achieves the low noise specifications with only 25% of the capacitance of a conventional solution.

Last but not least, the influence of cascaded and single-loop $\Delta\Sigma$ modulators on the phase noise has been investigated theoretically. A fast, non-linear analysis method is developed and verified that shows that the theoretical analysis is far too optimistic and that any non-linearity in the loop creates in-band noise leakage and spurious tones. The conversion of phase error to charge pump currents has been shown to be critical and an adequate design strategy is presented.

All this knowledge is united in a 2V, 1.8GHz $\Delta\Sigma$-controlled fractional-N synthesizer IC that includes a monolithic 35 kHz dual-path loop filter, an LC-tank VCO, a high-speed 64/79 prescaler, a zero-dead zone phase-frequency detector, a 3-step equalizer and dual charge pumps. The experimental results of the monolithic fractional-N synthesizer show compliance with the DCS-1800 standard and presents a highly-integrated solution with excellent spectral purity and fast dynamics in standard CMOS technology: the goal of the research. What is more, the same synthesizer is a key building block in one of the first monolithic CMOS transceiver front-ends that achieve cellular (DCS-1800) specifications.

List of Symbols and Abbreviations

Symbols

Physical

k	Boltzmann constant (1.38×10^{-23} (J/K))
q	Elementary charge (1.6×10^{-19} (C))
T	Absolute temperature
μ_0	Magnetic permeability of free space ($4\pi \times 10^{-7}$ (H/m))

Definitions

α	Zero positioning factor in PLL
α_{qp}	On-time fraction of charge pumps
β	Pole positioning factor in PLL
γ	Excess noise factor
	Fourth pole positioning factor in PLL
γ_{1/f^3}	$1/f^3$ phase noise corner frequency factor
$\Gamma(\omega_0 t)$	Impulse sensitivity function
Γ_{rms}	White noise conversion factor
$\Gamma_{1/f}$	Flicker noise upconversion factor
δ	Skin depth
δ_{Al}, δ_{Cu}	Skin depth of Aluminium and Copper
Δ	Quantization step
ε	PLL settling accuracy
ε_{ox}	Permittivity of oxide
$\theta, \theta(t)$	Phase and time-varying phase
$\Delta\theta$	Phase change
$\Delta\theta \rightarrow I_{qp}$	Phase error-to-charge pump current conversion
θ_x	Output phase of building block x
θ_{err}	Phase error
$\theta_{err,ss}$	Steady-state phase error
θ_{out}	Output phase
θ_m	Peak phase deviation at ω_m
θ_n	Phase noise due to voltage noise u_n
θ_{ref}	Reference phase
Θ_n, Θ_p	Mobility degradation factor for NMOS and PMOS
Λ_n, Λ_p	Channel length modulation factor for NMOS and PMOS
μ_{mos}	Mobility of the minority carriers in a MOS transistor

$\Delta\omega$	Offset frequency
	Frequency step
$\Delta\omega_{acc}$	Frequency accuracy
$\Delta\omega_H$	Hold-in range
$\delta(\omega)$	Dirac impulse function
ω	Radial frequency [rad/s]
ω_0	Resonance frequency, oscillation frequency
ω_c, ω_n	Cross-over and natural frequency of the PLL
ω_m	Modulating (baseband) frequency
$\Delta\Phi_{rms}$	Root-mean-square phase error
ϕ	Clock of a flipflop
ρ	Resistivity
ρ_{sub}	Resistivity of bulk substrate
σ_{Al}, σ_{Cu}	Conductivity of Aluminium and Copper
$\sigma_{\Delta\tau}$	Timing error variance
$\Delta\tau$	Timing error
τ_x	Time constant of x
ξ	Maximal over minimal capacitance ratio of a varactor
ζ	Damping factor
$a(t)$	Amplitude (time-varying)
A	Amplitude
	Amplifier noise contribution factor
A_i	Amplitude of signal i
A_n	In-band PLL noise
A_L	Inductor area
AF	HSpice 1/f noise modelling factor
AM_n	Amplitude conversion factor for $n\omega_0$
b	Number of output bits
B	Charge pump current multiplication factor
c_i	Noise upconversion factors of the ISF
C	Capacitance
C_{ox}	Oxide capacitance
C_{sub}	Substrate Capacitance
d	Distance between bonding wires
	Duty cycle
di_x^2	Average noise current of element x
dv_x^2	Average noise voltage of element x
$\frac{dg}{dx}$	VCO small-signal transfer function
DR_x	Dynamic range of x
e_i	Time representation of noise
$e(t), e[i]$	Continuous and discrete time representation of the quantization error
e_{rms}	Root-mean-square of the quantization error
$E_x(f), E_x(z)$	Fourier and z-transform of e_x
E	Energy

f	Frequency
f_0	Resonance frequency, oscillation frequency
f_T	Transistor unity current gain frequency
f_c	PLL cross-over frequency
f_{in}, f_{out}	Input and output frequency
f_{nbw}	PLL noise bandwidth
f_{ref}	Reference frequency of a PLL
Δf	Frequency noise
Δf_n	Frequency noise in the PLL band
F	Excess noise factor of the VCO
$g(x)$	Time-invariant transfer of the active element in a VCO
g_m, G_M	Transconductance
g_o	Transistor small-signal output conductance
G	Conductance
$G(s)$	Forward transfer function
$G_{lpf}(s)$	Loop filter transfer function
$GH(s)$	Open loop transfer function
$H(s)$	Feedback transfer function
	Resonator transfer function
$H_{qn}(z)$	Noise transfer function of a $\Delta\Sigma$ modulator
I	Current
I_b	Bias current
I_{qp}	Charge pump current
k	Internal accuracy of a $\Delta\Sigma$ modulator
K	Digital input word
K_0	PLL forward gain $[s^{-1}]$
K_f	Frequency dependent constant
K_n, K_p	Transconductance parameter of NMOS and PMOS
K_{lpf}	Loop filter gain
K_{pd}	Phase detector gain factor $[V/rad]$
K_{vco}	VCO gain factor $[rad/Vs]$
KF	Spice $1/f$ noise factor
l	Length of bonding wire
L	Inductance
	MOS channel length
L_{eff}	Effective MOS channel length
$\mathcal{L}_x\{\Delta\omega\}$	Single-sided phase noise spectral density of x at offset $\Delta\omega$
m	Filter order
M	Mutual inductance
Mi	MOS transistor no. i
N	Integer division modulus
N_{frac}	Fractional division modulus
n	Number of turns of a spiral inductor
	Bulk modulation factor
	$\Delta\Sigma$ modulator order

	Fractional ($0 \leq n \leq 1$) division modulus
P	Power
PM_n	Phase conversion factor for $n\omega_b$
Q	Quality factor of the LC-tank
Q_L	Quality factor of the inductor
Q_{MOS}	Quality factor of a MOS varactor
r	Radius
r_b	Bipolar small-signal base resistance
R	Resistance
R_{eff}	Effective resistance of an LC-tank
R_l	Parasitic series resistance of an inductor
R_{sub}	Resistance of bulk substrate
$R_{\square,x}$	Sheet resistance of x
R_{yy}	Discrete time autocorrelation estimate of y
$S_\theta(s)$	Phase power spectral density
$S_{x,y}(s)$	Power spectral density of x to y
t	Thickness
$T_{x,y}(s)$	Closed loop transfer function of x to y
T_c	Circuit delay time
T_d	Delay time
T_x	Time period of signal x
T_0	Time period of the carrier signal
T_ε	PLL settling time
V	Voltage
V_A	Voltage amplitude
V_{DS}, V_{GS}	Drain-source and gate-source voltage
V_T, V_{Tn}, V_{Tp}	Threshold voltage
v_c, V_c, V_{ctrl}	VCO control voltage
v_n	Noise voltage
v_{pd}	Phase detector output voltage
w	Width of metal turns in an inductor
Y	Admittance
Z_{lpf}	Charge pump loop filter impedance

Abbreviations

2G	Second Generation of wireless communication systems
2.5G	Transition from Second to Third Generation
3G	Third Generation
AC	Alternating Current
ADC	Analog-to-Digital Converter
AGC	Automatic Gain Control
ALAP	As Low As Possible
AM	Amplitude Modulation
AMPS	American Mobile Phone System

ASIC	Application Specific Integrated Circuit
AT&T	American Telephone and Telegraph company
BER	Bit Error Rate
BiCMOS	Bipolar and CMOS
BJT	Bipolar Junction Transistor
BTS	Base Transceiver Station
BW	Bandwidth
CDMA	Code-Division Multiple Access
CML	Current Mode Logic
CMOS	Complementary Metal Oxide Semiconductor
DAC	Digital-to-Analog converter
DC	Direct Current
DCS	Digital Cellular System
DDFS	Direct Digital Frequency Synthesis
DFF	D-FlipFlop
DMP	Dual-Modulus Prescaler
DPC	Digital-to-Phase Converter
DRO	Dielectric Resonance Oscillator
DSP	Digital Signal Processor
DSTC	Differential Single-Transistor-Clocked
DUT	Device Under Test
EDGE	Enhanced Data for Global Evolution
ETSI	European Telecommunications Standards Institute
EXOR	Exclusive OR
FDMA	Frequency-Division Multiple Access
FFT	Fast Fourier Transform
GaAs	Gallium Arsenide
GMSK	Gaussian Mean Shift Keying
GPRS	General Packet Radio Service
GPS	Global Positioning System
GSM	Groupe Spécial Mobile / Global System for Mobile communications
HF	High Frequency
HIPERLAN	HIgh PErformance Radio Local Area Network
HSCSD	High-Speed Circuit-Switched Data
IC	Integrated Circuit
IF	Intermediate Frequency
ISF	Impulse Sensitivity Function
IMT	International Mobile Communications
I/O	Input/Output
ITRS	International Technology Roadmap for Semiconductors
LDD	Lightly Doped Drain
LED	Light Emitting Diode
LF	Low Frequency
LNA	Low Noise Amplifier
LO	Local Oscillator

LPF	Low-Pass filter
LSB	Least Significant Bit
LTI	Linear Time Invariant
LTV	Linear Time Variant
MASH	Multi-Stage Noise-Shaping
MIM	Metal-Insulator-Metal
MS	Mobile Station
MSB	Most Significant Bit
NAND	Negative AND
NMOS	N-channel Metal Oxide Semiconductor
NTF	Noise Transfer Function
OSR	Oversampling Ratio
PCB	Printed Circuit Board
PCS	Personal Communications Service
PD	Phase Detector
PDC	Pacific Digital Cellular
PFD	Phase-Frequency Detector
PLL	Phase-Locked Loop
PM	Phase Modulation
	Phase Margin
PMOS	P-channel Metal Oxide Semiconductor
PSD	Power Spectral Density
PSK	Phase Shift Keying
RF	Radio Frequency
RX	Receive
SAW	Surface Acoustic Wave
SiGe	Silicon Germanium
SiP	System-in-Package
SNR	Signal-to-Noise Ratio
SoC	System-on-Chip
SSB	Single SideBand
TCXO	Temperature-Controlled Crystal Oscillator
TDMA	Time-Division Multiple Access
TSPC	True Single-Phase Clock
TX	Transmit
UMTS	Universal Mobile Telecommunications Service
VCO	Voltage-Controlled Oscillator
VCXO	Voltage-Controlled Crystal Oscillator
VFSR	Very Fast Simulated Re-Annealing
VGA	Variable Gain Amplifier
W-CDMA	Wideband Code-Division Multiple Access
WLAN	Wireless Local Area Network
xMOS	NMOS or PMOS transistor
Xtal	Crystal
YIG	Yttrium Iron Garnite

Contents

Chapter 1

Introduction

1.1 Telecommunications: An Overview

The dusk of the 20th and the dawn of 21st century will be remembered for the unrivaled growth of mobile telecommunications. Driven by synergetic economical and technological forces, small and cheap handheld devices have flooded the market, converting the mobile phone from a minority life-style choice to an ubiquitous possession. For example, the market penetration of cellular phones in Belgium has risen from 54% in 2000 to 79% in 2002. The same forces provided integrated circuit engineers with almost limitless availability of transistors, whose incremental cost nowadays is zero. But it wasn't always like that... The first section of this book is dedicated to the RF pioneers, who managed to squeeze remarkable circuit performance out of very limited resources.

It all started with the father of electromagnetics: James Clerk Maxwell. Not only did he formulate the electrodynamic equations in 1873, he also developed the first mathematical treatment of feedback control systems. Moreover, he laid the foundations for communications systems by inventing the displacement current, enabling the mathematical description of electromagnetic wave propagation. Heinrich Hertz was the first to verify experimentally Maxwell's theory by building spark gap "transmitters" and "receivers". These contraptions operated at 50-500 MHz(!) due to the limited size of his lab, but used vast amounts of power to produce a visible spark. The first real step towards wireless communications was made by Marconi; Starting from Hertz' transmitter and Branly's coherer [Lee98b, p.3], he managed to achieve transatlantic wireless communication in 1901. The search was on for an improved receiver, which resulted in the point contact crystal detector developed in 1904 by Bose [Lee98b, p.5]; This device is in fact a *semiconductor diode* (the diode symbol actually originates from this device). Because of the broad spectrum of the spark, a more selective CW detector was needed to reduce interference when the number of transmitters increased. This lead to the first AM voice and music broadcast by Fessenden on Christmas eve, 1906. 1907 was a breakthrough year with the invention of the first silicon detector (40 years before the invention of the transistor!), the first LED and the first triode vacuum tube, serendipitously discovered by de Forest, who called it the *audion*. After further improvements, the vacuum tube was ready to be used by clever engineers. In 1912, Edwin

Howard Armstrong invented the regenerative amplifier/detector [Arm15]. This circuit combined high gain and high Q extracted from one tube. Moreover, the tube non-linearity demodulated the signal. Additionally, when used with feedback, the device became the first RF oscillator. As if this wasn't enough, Armstrong invented the superheterodyne receiver in 1917, which still is the most commonly used receiver architecture. From the 1920s, developments in radio electronics accelerated and the communication age took off. A striking story is the one of the Soviet engineer Oleg Losev, who developed the first solid-state receiver in 1922. Starting from ZnO diodes, he constructed fully solid-state "crystadine" RF amplifiers, detectors and oscillators at frequencies up to 5 MHz, 25 years before the invention of the transistor!

The first wide-spreading telecommunication system was the telegraph, developed by Samuel Morse in 1837. In 1876, Graham Bell took wireline communications one step further by inventing the telephone for voice communication. Already one year later, the first commercial wireline telephone service was introduced by Bell Systems. In contrast, the first commercial mobile telephone service, developed by AT&T, saw the light of day in 1946. It was simple, oversized and could only handle a limited number of users. Bell Labs came up with the solution in 1947: *the cellular concept* ; The total area was divided in cells and communication channels could be passed on using "hand-off" procedures. While the first handheld cell phone was developed by Motorola in 1973 (see Fig. 1.2 (b)), the first commercial cellular system (AMPS) in the US began operation only in October 1983, due to years of political fights. The 36 years delay between the idea and the commercial exploitation are also due to the development of the key technology, with as milestones the invention of the transistor in 1947 (Bell Labs), the silicon transistor in 1954 and the integrated circuit in 1958 (both Texas Instruments). This enabled miniaturization of the mobile phone and microprocessor implementations to realize a smooth hand-off function. Meanwhile, in Scandinavia the Nordic Mobile Telephone (formed in 1969) came up with the first multinational cellular telephony system (450 MHz) in 1981. From then on, more countries followed, each with their own standard.

The increasing number of users and the need for roaming between countries brought up the demand for a standardized system with more efficient use of the available frequency spectrum. As early as 1978, the Conférence Européenne des Postes et Télécommunications (CEPT) reserved two blocks of 25 MHz in the 900 MHz band for mobile telecommunications in Europe. In 1982 a standardization working group was set up: the Groupe Spécial Mobile (GSM), which was transferred in 1989 to the European Telecommunications Standards Institute (ETSI).

The uprise of standardized, high-quality, multi-user telecommunication systems was catalyzed by the rapid technological evolution, which enables vast digital signal processing for efficient modulation schemes, processing and protocols. These so-called digital telecommunications system can be categorized as follows: GSM (a.k.a. Global System for Mobile communications) with three different flavors: GSM-900, DCS-1800 (GSM-1800), PCS-1900 (GSM-1900), which account for around 70% of the digital cellular market and CDMA and D-AMPS in the US and PDS in Japan, which take up the remaining 30%.

The technological evolution combined with economical motives pushes telecommunications standards to higher performances, i.e. to 3G (third generation) communications systems. 3G enables high-speed data communication, mobile Internet and e-commerce, video-on-demand, ... As a stepping stone to 3G, 2.5G standards have emerged that with little investment can boost

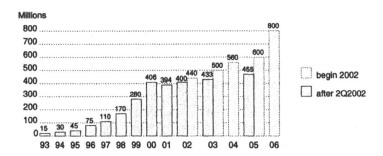

Figure 1.1:   *Cellphone sales in units over the years and in the future in the beginning of 2002 [Sil02, Future Horizons, UK] and after 2Q2002.*

the performance of classical GSM, e.g. GPRS (General-Packet-Radio Service) with data rates of 171.2 kbps and the newest high data rate GSM flavor, the Enhanced Data rate for GSM Evolution (EDGE) standard, with data rates up to 400 kbps. Eventually, 3G is expected to take over, with several standards: CMDAOne/CDMA2000, GSM/W-CDMA,TD-CDMA (Time Division CDMA) and UMTS (Universal Mobile Telecommunications System). In 2001, the first UMTS network was launched in Japan by NTT DoCoMo.

In parallel, short range RF networks are emerging. The Bluetooth™, Home RF and DECT standards belong to the low data-rate types with data rates around 1 Mbps at 2.4GHz. At the high-data-rate end WLAN (Wireless Local Area Network) are situated with standards as IEEE802.11a and its European sibling, HIPERLAN, both targeted for 54 Mbps at 5GHz and IEEE802.11b (11 Mbps at 2.4 GHz) and the new IEEE802.11g with enhanced data rates between 20-50 Mbps at 2.4 GHz. This abundance of wireless standards and communication types together with GPS, the single-chip cellular phone, the multi-standard phone and eventually 4G communication systems ensure the job-potential of telecommunications design engineers, now and in the future.

1.2   Telecommunications: A Market Perception

This book and the vast majority of all material published on RF circuits starts with: "The explosive, unrivaled, enormous (and all other possible superlatives) growth of mobile communications...". While this was true at the end of the twentieth century, with a compound growth rate of the number of handsets of 42.4% from 1995 to 2000 [Sil02, Future Horizons, UK], 2001 proved to be a disastrous year for the mobile communications industry and the IC industry in general; The global handset sales even *dropped* by 1% from 406 million to 394 million units in 2001, due to the reduced demand, inventory corrections and the resulting price drops for baseband and RF integrated circuits [Sil02, Frost&Sullivan]. The chip revenue per handset dropped from about $41 in 1999 to $31 in 2001, due to falling flash memory prices, increased integration, economies of scale and competition [Sil02, Future Horizons, UK].

However, forecasts in the beginning of 2002 look a lot more promising; Analysts predict an increase in handset sale of around 10% (some even 20%) [Sil02] to around 440 million handsets up to 800 million in 2006. Fig. 1.1 shows the cellphone sales curve over the years up to 2006 [Sil02, Future Horizons, UK]. The curve is nearly perfect apart from the hiccup in 2000. The growth is attributed to new cellular phone services and applications –Nokia plans 20 new products–, the growth in Asia (especially Korea and China) and of course the emergence of 3G. Also replacements determine the growth rate; It is estimated that around 35% of the handsets in 2002 are replacements and the number is expected to rise to 77% in 2006 [Sil02, Strategy Analytics]. Unfortunately, the economic situation did not improve as expected (due to e.g. September 11, 2001) and handset sale forecasts were tuned after 2Q2002 to a little more than 400 million for 2002 (by Nokia and Motorola) and around 433 million in 2003, with a worst-case forecast of only 455 million in 2005 (see Fig. 1.1).

Also the future of 3G has become less rosy then was predicted in the beginning of 2002. Scepticism exists since the cost to the user could be high because of the high investments, the smaller cells and the licenses. In addition, the in-the-field data throughput of 3G is around 56 kbps due to interference, such that 3G makes little difference to the normal consumer compared to the 2(.5)G networks. This is reflected in the limited success of the pioneering UMTS launch in Japan in 2001. NTT DoCoMo had to cut its subscriber forecast for the end of 2002 to 400,000, i.e. a third of its original predictions. On top of that, WLAN could gain the upper hand on 3G when WLAN "hot spots" or access points are set up all over the world in public areas, providing much higher data rates than 3G. At this moment (October 2002) the mobile phone industry is split between those who are under legal obligation to build new and costly 3G networks in the next few years (Europe) and those who can wait for consumer demand to justify the cost (Asia and the Americas). The latter don't foresee a need for 3G for another 4 to 5 years, while Europe expects the high-speed UMTS mobile internet market to take off before the end of 2004. As an engineer, I tend to think that 3G will make its way because of my belief in technology and people. People will buy these phones; Before the widespread use of 2G cellular phones, no one needed them, now everyone wonders how they managed without them. Whatever happens, the future of 3G lies in the hands of the mass consumer market...

1.3 Integration: Why, How and In What?

The synergism between economics and technological evolution, in hardware and software, drives phone manufacturers to higher levels of integration; In fact, the cost and ease of implementation of handsets becomes more critical, requiring aggressive reduction of the number of components. At the same time, the handset size is reduced; Mobile phones have evolved from bulky, expensive things (see Fig. 1.2) towards small, affordable and user-friendly handsets (see your pocket). This evolution continues ultimately towards the single-chip mobile phone (RF-SoC) or the single-package mobile phone (RF-SiP). Of course, integration requires architectural and technological innovations; The RF-front-end will move from superheterodyne to direct conversion architectures, with in 2006 direct conversion in over 95% of the handsets. Technology already moved from GaAs to silicon and is moving towards the most cost-effective solution: CMOS-RF. Disbelievers claim that CMOS-RF will only be chosen for low-end, low-cost systems, while the much more powerful SiGe-BiCMOS technology remains the choice for high-end systems. The aim of

Figure 1.2: *(a) The first car "mobile" phone in 1924. (b) The first handheld cell phone in 1973.*

this book is to prove them wrong and demonstrate the capabilities of high-end RF-CMOS.

1.3.1 P-Words

The reasons for integration are best summarized with the 4 *P-words*: *Performance-Power-Package-Price*, in rising order of importance. Of course, performance is a "conditio sine qua non". The benefits of integration for performance and power are less obvious and application dependent. Performance can be enhanced since the absence of external nodes reduces interference, noise and signal pick-up and avoids performance degradation by packaging parasitics and 50Ω terminations. Power consumption is possibly increased, since performance has to be extracted from the integrated circuits and not from the high-quality components. However, the higher predictability of integrated designs allows better optimization of the overall power consumption, such that the power advantage depends on the application and the quality of the design. Integration reduces the pin number and package size, leading to shrinking handsets. But the main reason for integration remains the price.

In the foreseeable future, market penetration will saturate and sales will slow down; The growth market will become a replacement market (77% in 2006), which is much more sensitive to economic fluctuation. This combined with the harsh competition in the mobile market, puts severe pressure on the margin and profit growth. The mobile manufacturer is therefore forced to reduce the cost to safeguard his profit. One way is to reduce the cost of labour by moving to low-wage countries, while the other way is to lower the cost of the electronic system. Therefore technology must shift from conventional discrete components and board-level assembly to chip- or package-level system integration. To reduce the cost, all tuning, trimming and postprocessing steps must be avoided.

It is clear that board level solutions are no longer acceptable. The ultimate goal is full RF Systems-on-Chip (RF-SoC) , although some reservation might be appropriate. First, due to the high complexity of RF-SoC, system reliability is critical requiring extensive error testing and flawless system and circuit verification, which increase the turnaround time of the designs. Sec-

Figure 1.3: *The interior of a modern mobile phone.*

ond, in spite of technology scaling, there are limits in lowering the power consumption and size of the SoC; Size reduces every technology node, but the non-scalable parts of the system, i.e. the passive components, the analog transistors (for matching,...) and I/O transistors, limit the power consumption reduction and *increase* the chip cost; It is estimated that the chip cost, while decreasing for 0.25μm and 0.18μm CMOS, increases for 0.13μm technologies (chip size is not much scaled but the wafer price is $\pm30\%$ higher) [Mat02]. Another integration solution is "System-in-Package"(SiP) ; SiP allows multiple technologies in single package, resulting in a more optimized solution with cost scaling along with technology scaling and more flexibility. It is believed that both RF-SoCs as SiPs will be developed, depending on the application; For a 0.35μm CMOS technology, SoC is for instance a more cost-effective solution [Mat02].

1.3.2 Direct Conversion Transceivers

The electronics of a mobile phone consists of roughly 2 parts (see Fig. 1.3): the radio-frequency (RF) and intermediate-frequency (IF) analog front-end (for transmission and reception) and the baseband and digital circuitry (for (de)modulation and intelligence). The back-end of modern mobile phone is already highly-integrated (more than is the case in Fig. 1.3). For the RF front-end most mobile phones use the now 85 year old superheterodyne receiver topology; The superheterodyne receiver thanks its performance to high-quality external SAW filters for interstage image filtering and IF selection. Although very potent, the extra components take up board space and increase the cost.

As pointed out in the previous section, integration is a must. Forecasts indicate that by 2005 direct conversion receiver will be used in 95% of all GSM cellular handsets up from 40% in 2001. Switching to direct conversion in GSM handsets reduces component costs as much as 30%, and 70% in CDMA cellular handsets [Sil02, iSuppli].

Different flavors of direct conversion architectures exist in the receive as well as the transmit path. Direct conversion receivers are zero-IF [Sev91] and low-IF receivers [Crol95] . Both architectures feature single-stage, dual-path, quadrature downconversion, without high Q filters. The architecture also fully exploits the core competence of CMOS by assigning the image rejection, channel selection and demodulation to the digital domain. The bottleneck of zero-IF receivers is $1/f$ noise and mismatch-related or self-mixing related offsets that corrupt the wanted signal at low frequencies. The low-IF receiver alleviates this problem, by setting the IF frequency to

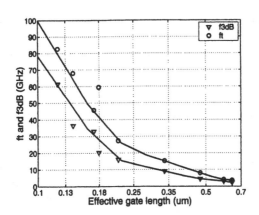

Figure 1.4: *The f_T and f_{3dB} of a NMOS transistor for different gate lengths and a $V_{GS} - V_T$ of 0.25V.*

typically half the channel bandwidth [Crol95]. But there is no substitute for cubic inches; Removal of the high Q, area expensive filters, imposes higher demands on the analog building block performance, and thus the power consumption.

Direct conversion transmitters are mostly Cartesian I/Q modulators, which are in fact the equivalent of the zero-IF receive architecture. Quadrature, single-path conversion alleviates the need for high Q filters. Another direct conversion transmitter that is emerging is the direct modulation of a fractional-N synthesizer by the digital data, which presents a elegant and highly-integrated solution [Perr97].

1.3.3 CMOS Technology

In the early 1980s, the GHz spectrum was GaAs territory. Since the cost of silicon wafers is much lower than that of GaAs, silicon technologies conquered the market in the late 80s. Typical NPN f_Ts were 10 GHz, sufficient for circuits operating at 1 GHz. Nowadays, the bipolar f_T is saturating around 40 GHz, but introduction of SiGe bipolar transistors improved this to 100 GHz, which is higher than classical GaAs FETs. In the total semiconductor revenue GaAs only takes up 1% while silicon (Bipolar and CMOS) takes up 99% [Mat02].

Today's high-end RF front-ends are mostly implemented in BiCMOS, SiGe, Bipolar and/or GaAs. However the cost of a CMOS wafer is much lower; CMOS can benefit from a high-volume market, which aggressively scales technology to improve the digital performance, which at the same time improves the RF performance. The ITRS road map [ITRS99] predicts a transition to another technology node every 3 years. In reality, this happens every 2 years! The conventional scale factor for one technology node is $1/\sqrt{2}$. The transistor area is scaled by a factor 1/2 while the integration density is doubled. Cost is scaled by $1/\sqrt{2}$, while the operating frequency is increased by $\sqrt{2}$. While CMOS was never considered an option for RF, the device scaling has led to MOS transistors with an f_T of almost 100 GHz. In Fig. 1.4, the f_T and f_{3dB} values are plotted for different in-house available technologies for an NMOS with a $V_{GS} - V_T$ of 0.25V. The

f_{3dB} gives a more realistic idea of the real frequency performance of a technology [Stey93], since it takes into account all gate and drain capacitances and not only the gate-source capacitance. But also the f_{3dB} is getting close to 100 GHz. While industry has maintained Moore's law for almost 30 year, some saturation in the scaling is expected (even by Moore himself) when fundamental borders are reached, e.g. gate oxide thicknesses at atomic level. But performance can still be increased by "equivalent" scaling, i.e. new materials like Cu interconnects, low-k dielectrics or more metal layers, all of which are at the same time beneficial for the RF performance of CMOS technology.

The main reason to consider CMOS for RF integration is of course the cost. It has been estimated that RF-CMOS increases the wafer cost compared to plain CMOS by around 15%, due to MIM capacitances, resistances, ... But SiGe-BiCMOS increases the wafer cost by more than 40% [Mat02]! In SiGe-BiCMOS performance with low power is more easily obtained, but because of the lack of competition in this industry compared to CMOS, the price is not likely to decrease. Together with the compatibility of CMOS with the digital back-end, CMOS is an attractive cost-effective solution for integration of RF front-ends, with as goal full RF-SoCs. The goal of this book is to prove the capability of RF-CMOS for high specification standards, such as DCS-1800 at a sensible power consumption.

1.3.4   Trends in Research Evolution

The trend towards higher degrees of integration in the most cost-effective technology can be retraced in the history of mobile transceiver, reconstructed by key publications. In the early years of mobile communications, all RF front-end ICs for cellular and cordless phones were integrated in Si bipolar. Nice examples are the transceiver front-ends of [Sev94, Kou93]. Already people thought of highly-integrated solutions, by using zero-IF architectures [Sev94].

In the following years publications emerged with transceiver building blocks in BiCMOS due to the continuing improvement of mainly the bipolar transistor. In 1995, the first GSM transceiver ICs were published: [Mars95] in BiCMOS and [Stet95] in Si bipolar, both with superheterodyne architectures with external SAW filters, LC-tank, loop filters.

In the same year CMOS made its entry in the GHz range with a 900 MHz low-IF receiver [Crol95] and a high performance 1.8 GHz VCO [Cran95a]. From then on research for RF-CMOS took off rapidly; Industry remained sceptical, but universities went into competition to earn credits in this field. The first CMOS papers were mainly focused on LNA, mixer and VCO building block design [Kara96, Rof96b].

In 1998, the first single-chip CMOS transceiver demonstrating the feasibility of achieving cellular specifications in CMOS was presented by the KULeuven [Stey98]. The IC was aimed for DCS-1800 and combined a highly-integrated low-IF receiver with a direct conversion transmitter and an on-chip VCO. Meanwhile, other research groups as well as industry focused on lower-end application for RF-CMOS with in the same year the presentation of a 900 MHz single-chip transceiver for the ISM band [Rof98a, Rof98b]. In 1999, [Cho99] demonstrated a single-chip transceiver for 900 MHz spread-spectrum cordless telephony, with a direct conversion receiver and transmitter. A 5 GHz WLAN CMOS chip-set with direct conversion in the TX and RX path was published by Bell Labs in 2000 [Liu00].

In 2000, the high-spec barrier was broken by CMOS; [Stey00a] presented a 2V CMOS cellular transceiver front-end for DCS-1800. This IC was the first single-chip CMOS transceiver front-end with a complete on-chip PLL (with VCO and loop filter) that achieved the requirements of the stringent DCS-1800 cellular system. Industry on the other hand, mainly sticks to BiCMOS and bipolar, but becomes slowly more convinced of the capabilities of RF-CMOS, which is reflected in the first commercial CMOS transceiver chip set for GSM [Aero01, Fren01].

In parallel, the integration of RF network transceivers in CMOS gained more and more attention with in 2001 a full session at the International Solid-State Circuit Conference (ISSCC) on WLAN featuring two CMOS Bluetooth transceivers [Ajj01, Dar01] and 1 CMOS Bluetooth SoC [Eynd01]. In 2002, more CMOS Bluetooth transceivers and a CMOS IEEE802.11a transceivers [Su02] were published and even 3G (UMTS [Man02] and WCDMA [Bren02]) transceiver blocks in CMOS were presented, proving the increasing interest and believe in RF-CMOS for research as well as commercial use.

1.4   The Research Book

The presented research fits in the quest for a fully integrated, CMOS-only transceiver front-end, that meets the stringent cellular specifications. The DCS-1800 cellular standard has been chosen as the driving application to demonstrate the cellular potential of RF-CMOS even for high frequency standards. As stated in the previous section, this quest was successful and led to the presentation of the first CMOS cellular transceiver front-end in 2000 [Stey00a, Stey00b], which meets the DCS-1800 specifications [ETSI00]. The IC features a fully integrated low-IF receiver [Jans02], a fully integrated direct-conversion transmitter [Borr02] and a monolithic $\Delta\Sigma$-controlled fractional-N frequency synthesizer, with an on-chip loop filter and a fully integrated high-quality VCO.

A major challenge in the design of CMOS single-chip transceiver systems is the frequency synthesizer. The spectral purity of the frequency synthesizer is the most critical parameter for the quality and reliability of the information transfer. In addition, the dynamic behavior of the frequency synthesizer determines the maximal information throughput, which is of utmost importance for 2.5G and 3G wireless communication systems. This book presents the design trajectory of a monolithic frequency synthesizer synthesizer in standard CMOS technology, focused on attaining the highest spectral purity and fastest switching with moderate power consumption. The design trajectory goes from the specification derivation down to the systematic development of different CMOS circuits, with ultimately the successful implementation of a monolithic frequency synthesizer. Along the path, CMOS frequency synthesizer design is investigated on all levels: from architectural level to building block level down to circuit level and back up.

The architectural level of the design flow is mainly application-driven and led to the choice of a $\Delta\Sigma$ fractional-N PLL architecture, since it trades off accuracy (i.e. high specifications) for digital complexity, i.e. the natural biotope of CMOS. By carefully interpreting the DCS-1800 requirements [ETSI00], measurable synthesizer specifications have been extracted, more specifically phase noise, rms phase error, spurious suppression and settling time. The relative importance of the different building blocks on the overall synthesizer performance has been investigated and used to set their requirements. The translation of the derived building block specifications onto actual circuits down to transistor level forms the more theoretical core of the presented book. Insights on PLL and PLL building block design are gathered and design guidelines are given.

- In the quest for the fastest divide-by-2 flipflop, a systematic sizing strategy is employed based on optimization of the circuit for divide-by-2 operation and careful design of each circuit node for speed and power. Moreover, design rules are developed to design the prescaler circuits for low residual phase noise and high input sensitivity.

- A small digression is made onto the exploration of the speed limits of conventional CMOS. A CMOS prescaler has been shown to operate at above-12 GHz, even in first-generation 0.25μm technologies.

- For VCO design, a design-oriented phase noise theory is presented that not only takes into account the noise of the passive and active devices, but also the influence of the non-linear active element. A simulator-optimizer program is developed to realize optimal inductor geometries in CMOS to reduce white phase noise generation.

- Moreover, 1/f noise upconversion mechanisms have been identified and successfully countered in the VCO circuit design, by choosing the operating point that maximizes waveform symmetry. A bias filtering technique to remove all flicker noise that also significantly reduces white phase noise has been elaborated and experimentally verified.

- Monolithic PLL design is theoretically analyzed for noise and capacitance; A systematic design strategy has been developed that marks out the design space, enabling optimized, fully integrated loop filters.

- The influence of $\Delta\Sigma$ modulator noise in fractional-N PLLs is theoretically derived and loop bandwidth limitations are elaborated for phase noise and rms phase error.

- A fast non-linear analysis method has been developed to more accurately predict the effect of $\Delta\Sigma$ modulators in fractional-N PLLs; Non-linearities, i.e. a dead zone and gain mismatch, in the PFD-charge-pumps are identified as the main sources of in- and out-of-band $\Delta\Sigma$ noise leakage and spurious tones. The fractional-N PLL has been found most sensitive to spurious tones near integer division moduli.

This research has led to several CMOS implementations of PLL building blocks and a fully integrated $\Delta\Sigma$ fractional-N frequency synthesizer.

- A single-ended 1.5 GHz 8/9 dual-modulus prescaler in 0.7μm CMOS with optimization of the residual phase noise and high input sensitivity [DeMu98].

- A single-ended 1.8 GHz 8/9 dual-modulus prescaler in 0.8μm BiCMOS with a custom ratio-ed logic D-flipflop to maximize the speed-power ratio with a high input sensitivity.

- A 12 GHz /128 prescaler in a first-generation 0.25μm CMOS which features a divide-by-2 flipflop, operating with small input signals up to 15 GHz [DeMu00c].

- A 2 GHz low-phase-noise integrated LC-VCO set in $0.65\mu m$ BiCMOS with circuit tricks to minimize the flicker noise upconversion. Phase noise is as low as -125 dBc/Hz at 600 kHz and the $1/f^3$ phase noise corner is reduced to 15 kHz [DeMu00b, DeMu99a]. In a small frequency band, the VCO has no flicker noise upconversion and phase noise as low as -132.5 dBc/Hz at 600 kHz due to a low-impedance bias current source filtering technique.

- A 1.8 GHz fully integrated VCO in $0.25\mu m$ CMOS with a tuning range of 28% and phase noise as low as -127.5 dBc/Hz at 600 kHz, which was the best ever reported CMOS VCO at that time in open literature [DeMu00a].

- A 2V, 1.8 GHz fully integrated $\Delta\Sigma$-controlled fractional-N PLL in $0.25\mu m$ CMOS. The IC includes a fully integrated 35 kHz dual-path loop filter, a fully integrated LC-tank VCO, a high-speed 64/79 prescaler, a zero-dead-zone phase-frequency detector, a 3-step equalizer and dual charge pumps. Its phase noise is as low as -125.7 dBc/Hz (integer) [DeMu00d, Stey00b] and below -120 dBc/Hz (fractional) at 600 kHz and the frequency resolution is 400 Hz, with only 1.4 nF on-chip [DeMu02].

The VCO of the presented fractional-N synthesizer is capable of driving a polyphase filter in a power efficient manner [Borr00]. These realizations pave the way to the integration of a complete transceiver in CMOS, without trimming, tuning and external components. The PLL operates from a power supply voltage of 2V to prove the feasibility of low-voltage RF-CMOS for compatibility with the digital part in view of single-chip transceiver integration.

1.5   The Outline of the Book

The outline of the presented book is as follows; Chapter 2 introduces the frequency synthesis basics. The DCS-1800 system is discussed and the requirements of the frequency synthesizer in the DCS-1800 system are derived from *BER* to spectral purity specifications.

Chapter 3 covers the design of dual- and multi-modulus prescalers going from architecture to the search for the best high-speed flipflop over different specifications: speed and power, but also phase noise and input sensitivity. The feasibility of plus-12 GHz CMOS prescalers is demonstrated. In Chapter 4, a design-oriented oscillator phase noise theory, which takes into account the noise conversion due to the non-linearity of the active element is presented. High-Q inductor design guidelines are given, leading to optimized inductors by an in-house developed simulator-optimizer program. Moreover, 1/f noise upconversion mechanisms have been identified and successfully countered in the circuit design. A VCO bias filtering technique has been experimentally verified that removes all flicker noise and seriously reduces white phase noise.

Chapter 5 goes up a level and elaborates the loop filter design for a 4th-order, type-II PLL, which trades off noise contributions of all building blocks for integrated capacitance. A dual-path loop filter is developed that achieves the noise specifications with only 25% of the capacitance necessary in a conventional 4th-order PLL.

Finally, in Chapter 6 the 1.8 GHz monolithic $\Delta\Sigma$ fractional-N synthesizer is presented. The influence of $\Delta\Sigma$ modulator phase noise has been investigated theoretically. A fast, non-linear analysis method is developed that shows that any non-linearity in the loop creates large in-band noise leakage and spurious tones. The PFD and charge pumps have been shown to be critical and an adequate design strategy is presented. The experimental results of the monolithic fractional-N synthesizer show compliance with the DCS-1800 spec and presents one of the most powerful, most integrated solutions published in open literature today.

To conclude, Chapter 7 gives an overview of the design of the 2V CMOS transceiver front-end and ends with concluding remarks.

Chapter 2

On Frequency Synthesis

2.1 Introduction

To constitute a complete transceiver for wireless communication systems, one indispensable building block is required by both the receive and transmit path, i.e. the building block that generates the local oscillator (LO) signal: *the frequency synthesizer*. Wherever frequencies are translated, frequency synthesis is crucial to provide a clean, stable and programmable local oscillator signal; It needs to be programmable to address all frequency channels, it needs to be fast switching to perform the addressing sufficiently fast and it needs to be clean, since low oscillator noise is vital for the quality and reliability of the information transfer.

The aim of this book is to implement a frequency synthesizer that satisfies the same constraint imposed on the receiver and transmitter: cost reduction. The same trends apply to a cost-effective synthesizer implementation: finding ways to go to highly integrated solutions and achieve this integration in the cheapest technology, which is currently standard digital CMOS.

Frequency synthesizers typically come in three main flavors: the table-look-up synthesizer, the direct synthesizer and the indirect or phase-locked loop synthesizer [Egan81]. In a table-look-up synthesizer or digital synthesizer, the sinusoidal waveform is created piece by piece by using the digital values of the waveform stored in a memory. The hardware needed is a digital accumulator, whose capacity determines the frequency resolution, a memory containing a cosine, a digital-to-analog converter (DAC) and a low-pass filter to remove high-frequency spurs. Due to the limited speed of the memory and the high-resolution DAC, high frequency operation is not feasible. Moreover, high-frequency spurious tones tend to corrupt the spectral purity. The direct synthesizer synthesizes the wanted output frequency from a single reference by multiplying, mixing and dividing. By repeatedly mixing and dividing any frequency accuracy is attainable. Ideally, the output spectrum is as clean as the reference spectrum and fast frequency hopping is possible. However, when implementing the direct synthesizer, cross-coupling between stages is a serious problem for the spectral purity and the large number of components causes the synthesizer to be very bulky and power hungry. The indirect frequency synthesizer generates its output by phase-locking the divided output to a reference signal. The phase-locked loop frequency synthesizer has the potential of combining high frequency and low power. But its most distinct advantage is that the phase-locked loop is very well suited for integration in low-cost IC processes, like CMOS. This is the reason why the PLL is used for frequency synthesis in almost all wireless communication chip sets on the market.

The phase-locked loop synthesizer is the main topic of this book. In the remainder of the chapter, phase-lock and its application in frequency synthesis in different forms is described. To demonstrate the feasibility of high-quality, monolithic frequency synthesizers in CMOS technology, the stringent specifications of the DCS-1800 standard – or GSM-1800 – are set as a design target; The DCS-1800 communication system and the intended front-end integration are elaborated. Next, the data sheet of the frequency synthesizer is examined, going from static performance – phase noise, spurious suppression and phase error– over dynamic performance –tracking, settling– to more general specs as power consumption and integratability. Using the DCS-1800 standard as a demonstration vehicle, general DCS-1800 requirements are mapped onto the frequency synthesizer specifications, which are then further projected on PLL building block specifications. The different existing implementations of the basic PLL building blocks are discussed, with each their influence on the overall frequency synthesizer performance.

2.2 Indirect or Phase-Locked Loop Frequency Synthesizers

2.2.1 Phase-Lock

However simple the principle of phase-lock might seem, it is omnipresent in daily life and of utmost importance to maintain order in almost any electronic and mechanical contraption devised by man. By turning on a radio, a phase-locked loop starts tracking your favorite radio station and makes sure the radio receiver stays locked at this frequency. In a television set, a phase-locked loop keeps heads at the top of the screen and feet at the bottom. In color television, another phase-locked loop synchronizes color bursts, ensuring that green is green and red is red. Other examples are countless.   In an attempt to generally define an electronic phase-locked loop, one can say it is a circuit that causes one system to track another; More precisely, a PLL is a circuit that synchronizes an output signal an input signal in frequency as well as in phase. In the synchronized or *locked* state, the phase error between the output signal and the input signal is zero or very small. If a phase error builds up, a control mechanisms redirects the output signal as to minimize the phase error with the input signal. The phase of the output signal is actually locked to the phase of the input signal hence a *phase-locked* loop.

A phase-locked loop consists of three (or optionally four) basic components (Fig. 2.1): a phase detector (PD), a loop filter, a voltage controlled oscillator (VCO) and optionally a frequency divider. The phase detector compares the phase of the input signal against the phase (or the divided phase) of the VCO. The output of the phase detector is a measure of the phase difference between the two inputs. The difference voltage is then filtered by the loop filter and applied to the VCO. The control voltage on the VCO changes the frequency in the direction that reduces the phase difference between the input signal and the local oscillator.

The very first phase-locked loop was implemented in 1932 by de Bellescize [Bel32], a.k.a. the inventor of *coherent communication*. Superheterodyne receivers had come into use during

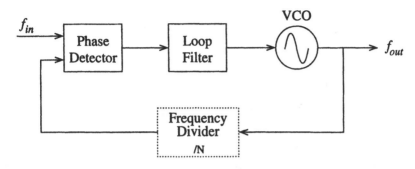

Figure 2.1: *The basic Phase-locked Loop.*

the 1920s, but the search for simpler techniques brought researchers to the synchronous receiver. The synchronous receiver basically consists of an oscillator, a mixer and an amplifier. To accurately receive an incoming signal the oscillator must be at the same frequency and phase, i.e. phase-locked by a phase-locked loop. In spite of its simplicity, the synchronous receiver never attained the popularity of the superheterodyne receiver. The widespread use of phase-locked loops started with synchronization of the horizontal and vertical scan in television receivers. At first, synchronization was performed by two free-running relaxation oscillators which act as sweep generators. The oscillators were synchronized at the start each line and at the start of each interlaced half-frame, which was marked by synchronization pulses transmitted with the video information. However, in the presence of noise, start-jitter occurred resulting in horizontal black streaks and vertical movement of the picture. To enhance the quality of the TV receiver, the oscillators were phase-locked to the sync pulses, by so called "flywheel" synchronizers implemented by phase-locked loops. The term "flywheel" originates from the high inertia, i.e. the low bandwidth, of the synchronizer, enabling it to move through periods of high noise and weak input signal without loosing synchronization. A third stepping stone in the uprise of phase-lock is space flight. The first American satellites communicated with earth using a low-power (10mW) CW transmitter, resulting in a weak received signal. Due to the Doppler effect and drift in the transmitting oscillator, there was considerable uncertainty on the received frequency. A fixed receiver would need a high bandwidth to receive the inherent narrow band signal, resulting in serious degradation of the SNR. The solution to this was once again a receiver that could lock onto the signal and reject lots of noise: a narrow band phase-locked loop.

The popularity of phase-locked loops originates from its use as a signal detector, i.e. the detection of a low-level signal carrying information in its phase or frequency, embedded in lots of noise. The task of the phase-lock receiver is to reproduce the original signal while removing as much of the noise as possible. To suppress the noise, the phase error voltage at the output of the phase detector is averaged over time and the average establishes the output frequency of the VCO. The longer the time averaging, i.e. the narrower the loop filter bandwidth, the more noise is eliminated.

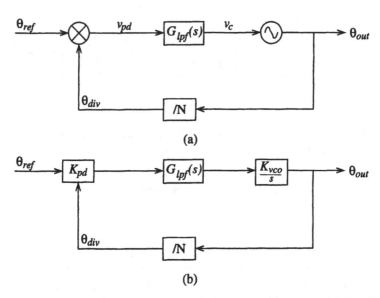

Figure 2.2: *(a) The basic phase-locked loop block diagram with state variables. (b) The basic linear time-invariant PLL model.*

2.2.2   Phase-Lock Frequency Synthesis Fundamentals

However, if the input to the PLL is a signal with a high signal-to-noise ratio, a noisy oscillator can be enclosed in the loop. The oscillator tracks out its own noise and its output is greatly cleaned up. The PLL now acts as a frequency synthesizer by locking the output frequency to the very clean reference input or a multiple of it, depending on the frequency divider (see Fig. 2.1).

The output frequency generated by the VCO is divided by a number N in the frequency divider. The phase of the divided frequency is compared in the phase detector to the phase of the clean reference frequency $f_{in} = f_{ref}$, generated by a Xtal oscillator. The voltage output signal of the phase detector is proportional to the phase difference between both inputs and consists of a AC and a DC component. The DC component is accumulated and the AC component is filtered by the low-pass loop filter. The resulting signal is a DC signal with a small superimposed AC signal, which serves as the control voltage for the VCO. When the phase relationship between the inputs is constant, the PLL is in lock and the output frequency equals :

$$f_{out} = N \cdot f_{ref} \qquad (2.1)$$

Although the phase-locked loop is inherently a non-linear feedback system, due to the non-linear PD, it can be accurately modeled as a linear time-invariant system when in lock, assuming that the PD transfer characteristic is linear in this operating region. Since it is a phase-locked loop, the input and output variables of the PLL are phases. In Fig. 2.2 (a), the basic block diagram of the PLL with its state variables is shown.

The reference signal has a phase θ_{ref} and the VCO output has a phase θ_{out}. The frequency divider divides the VCO frequency, and thus the VCO phase, by N i.e. $\theta_{div} = \theta_{out}/N$. When in lock, the output of the PD is a voltage proportional to the phase difference between the inputs:

$$v_{pd} = K_{pd} \cdot (\theta_{ref} - \theta_{div}) \tag{2.2}$$

with K_{pd} the phase detector gain in volts per radian. The phase error voltage v_{pd} is filtered by the loop filter, which normally has low-pass transfer characteristic. Noise and high frequency signals, i.e. the unwanted AC components, are suppressed; In addition, the loop filter determines the dynamic performance of the loop. The loop filter transfer function is given by $G_{lpf}(s)$.

The VCO frequency is determined by the control voltage v_c. Frequency changes due to the control voltage are given by $\Delta \omega = K_{vco} \cdot v_c$ where K_{vco} is the VCO gain and has units of rad/Vs. Since the control voltage acts on the frequency, which is the derivative of the phase, the VCO operation is modeled by $d\theta_{out}/dt = K_{vco} \cdot v_c$. By taking the Laplace transform this becomes

$$\theta_{out}(s) = \frac{K_{vco} \cdot v_c(s)}{s} \tag{2.3}$$

The linear time-invariant model of the PLL is drawn in Fig. 2.2 (b). The linear model of the PLL can be viewed as a standard feedback system with a forward transfer function $G(s) = K_{pd} \cdot G_{lpf} \cdot K_{vco}/s$ and a feedback gain, $H(s) = 1/N$. The open loop transfer function is then

$$GH(s) = \frac{K_{pd} \cdot G_{lpf} \cdot K_{vco}}{N \cdot s} \tag{2.4}$$

This transfer function determines the static and dynamic performance of the loop. It also reveals the major issue in PLL design: the loop bandwidth trade-off; For suppression of the unwanted AC components, the loop bandwidth needs to be as narrow as possible, while for the dynamic performance the opposite is true.

2.3 The Synthesizer Data Sheet

2.3.1 Spectral Purity

2.3.1.1 Definition of Phase Noise

Phase noise is a measure for the spectral purity of a signal and is of utmost importance in communication systems; Phase noise degrades the quality of TV pictures, limits the precision of satellite positioning and spoils the quality of data transmission/reception.

The output of an ideal oscillator is a perfect sinusoidal wave of frequency ω_0, i.e. $V_{out}(t) = A \cdot \sin(\omega_0 t + \theta)$ with A the amplitude and θ a fixed phase reference. In the frequency domain this corresponds to a Dirac impulse at ω_0, $\delta(\omega_0)$. In a real oscillator, noise generates fluctuation on the phase and the amplitude of the signal. The output signal becomes

$$V_{out}(t) = (1 + a(t)) \cdot \sin(\omega_0 t + \theta(t)) \tag{2.5}$$

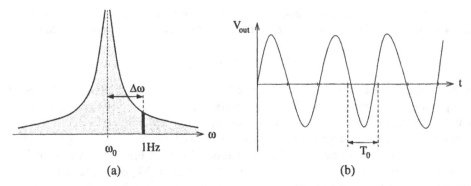

Figure 2.3: *The frequency (a) and time (jitter) (b) representation of phase noise in an oscillator.*

Due to the fluctuations on phase and amplitude, $\theta(t)$ and $a(t)$, the output spectrum is no longer a Dirac impulse, but exhibits sidebands close to the oscillator frequency, as shown in Fig. 2.3 (a). Since well-designed, high-quality oscillators are usually very amplitude stable, $a(t)$ can be considered constant over time. To quantify phase noise $\theta(t)$, the noise power in a unit bandwidth at a certain offset frequency $\Delta\omega$ from ω_0 is considered and divided by the carrier power. The result is a single sided spectral noise density in units dBc/Hz:

$$\mathcal{L}\{\Delta\omega\} = 10\log\left(\frac{\text{noise power in a 1 Hz band at } \omega_0 + \Delta\omega}{\text{carrier power}}\right) \qquad (2.6)$$

The question remains how noise on the phase of a sinusoidal signal translates to a phase noise skirt in the oscillator's output spectrum. Suppose that the phase fluctuation in Eq. (2.5) is a single sinusoidal tone in the phase, $\theta(t) = \theta_m \cdot \sin(\omega_m t)$ with $\theta_m \ll 1$, then the output of the oscillator becomes:

$$V_{out}(t) \approx A \cdot \sin(\omega_0 t) + A \cdot \frac{\theta_m}{2} \cdot \left[\sin\left((\omega_0 + \omega_m)t\right) + \sin\left((\omega_0 - \omega_m)t\right)\right] \qquad (2.7)$$

The output spectrum of the oscillator consists of a narrow-band FM signal with θ_m the modulation index, resulting in a strong component at ω_0 and two small "side lobes" at $\omega_0 \pm \omega_m$. Therefore, the oscillator output voltage power spectral density (PSD) is directly related to the phase noise PSD.

$$S_\theta(\omega) = \frac{\theta_m^2}{2} \cdot \delta(\omega - \omega_m) \qquad (2.8)$$

$$S_{V_{out}}(\omega) = \frac{A^2}{2} \cdot \left[\delta(\omega - \omega_0) + \frac{1}{2}S_\theta(\omega - \omega_0) + \frac{1}{2}S_\theta(\omega_0 - \omega)\right] \qquad (2.9)$$

Since the phase noise spectrum can be seen as a sum of sines, the total phase noise skirt is directly translated to noise side lobes at both sides of the carrier frequency (see Fig. 2.4). The actual phase noise at an offset ω_m is then:

$$\mathcal{L}\{\omega_m\} = 10\log\left(\frac{S_{V_{out}}(\omega_0 + \omega_m)}{A^2/2}\right) = 10\log\left(\frac{S_\theta(\omega_m)}{2}\right) \qquad (2.10)$$

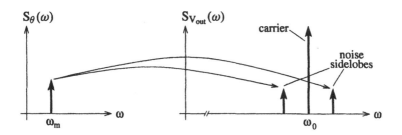

Figure 2.4: *From noise on the phase to the oscillator output spectrum.*

Phase noise is related to the PSD of instantaneous frequency deviations $\Delta f(t)$, since frequency is the derivative of phase:

$$S_{\Delta f}(\omega) = \omega^2 \cdot S_\theta(\omega) = 2\omega^2 \cdot 10^{\mathcal{L}\{\omega\}/10} \qquad (2.11)$$

The noise seen in the frequency spectrum has also an effect in the time domain. Due to the noise related phase/frequency deviations, there is uncertainty on the zero-crossings of the oscillator signal; the exact time of one period of the sine wave differs from period to period (see Fig. 2.3 (b)). Since θ_m is small, the waveform is almost-periodic, with an average period of T_0, corresponding to ω_0 and a timing error $\Delta\tau$. Several measures for the timing error exist such as *cycle-to-cycle* jitter, i.e. the difference between two consecutive periods, or as *absolute* jitter, i.e. the timing error with a periodic waveform of the same frequency. Since the phase noise is considered stationary, ergodic and Gaussian, the standard deviation or effective variance, $\sigma_{\Delta\tau}$ is the rms value of the timing error. A first-order formula to relate jitter to white phase noise can be derived from the more general definition for jitter, the *Allan variance* [Rhod83, p. 73], with T_0 the period of the oscillator signal:

$$\mathcal{L}\{\omega_m\} = 10\log\left[\frac{2\pi\,\omega_0}{\omega_m^2} \cdot \left(\frac{\sigma_{\Delta\tau}}{T_0}\right)^2\right] \qquad (2.12)$$

As mentioned before, the phase noise limits the precision of different communication systems. Typically, in a communication system a small wanted signal must be detected in the presence of large unwanted signals. In the receive path, the wanted signal is downconverted by the local oscillator signal. If a noisy oscillator is used, the large unwanted signal is downconverted by the phase noise at the offset frequency of the unwanted signal, as illustrated in Fig. 2.5; The dark box represents unwanted noise in the wanted signal band. If the phase noise at the given offset frequency is not sufficiently low, the unwanted noise in the signal band seriously degrades the signal-to-noise ratio. In the transmit path, the high-power output signal must satisfy a given power mask, such that the transmitted noise does not drown other transmitted signals. Therefore, the output spectrum of the oscillator must comply with the same power mask, such that the up-converted signal does too. Usually, this requirement is much less stringent than the one imposed by the reception quality.

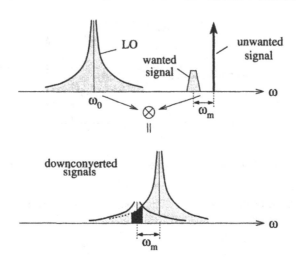

Figure 2.5: *The effect of phase noise on the reception quality of a communication system.*

2.3.1.2   Phase Noise in a PLL

Using the linear PLL model defined in Fig. 2.2 (b), the transfer function for different noise sources within the PLL to the output can be derived. The two most important source are noise coming from the reference signal, θ_{ref} and the phase noise of the VCO, θ_{vco}. When the VCO phase noise is modeled as an additive noise source behind the VCO block, the closed-loop response to VCO noise becomes:

$$\frac{\theta_{out}(s)}{\theta_{vco}(s)} = \frac{1}{1 + GH(s)} = \frac{sN}{sN + K_{pd} \cdot G_{lpf}(s) \cdot K_{vco}} \tag{2.13}$$

In the same way the closed-loop response to reference noise is

$$\frac{\theta_{out}(s)}{\theta_{ref}(s)} = \frac{G(s)}{1 + GH(s)} = \frac{N \cdot K_{pd} \cdot G_{lpf}(s) \cdot K_{vco}}{sN + K_{pd} \cdot G_{lpf}(s) \cdot K_{vco}} \tag{2.14}$$

To illustrate how both noise contributions are transferred to the output, the most elementary loop filter is chosen, i.e. $G_{lpf}(s) = K_{lpf}$. The open loop transfer function is than

$$GH(s) = \frac{K_{pd} \cdot K_{lpf} \cdot K_{vco}}{sN} = \frac{K_o}{sN} \tag{2.15}$$

with K_o the forward gain of the PLL in s^{-1}.

Using Eq. (2.13) and Eq. (2.14), the closed loop transfer function for a constant loop filter can be calculated:

$$\frac{\theta_{out}(s)}{\theta_{vco}(s)} = \frac{1}{1 + K_o/(sN)} = \frac{s}{s + \omega_c} \tag{2.16}$$

$$\frac{\theta_{out}(s)}{\theta_{ref}(s)} = \frac{K_o/(sN)}{1 + K_o/(sN)} = N \cdot \frac{\omega_c}{s + \omega_c} \tag{2.17}$$

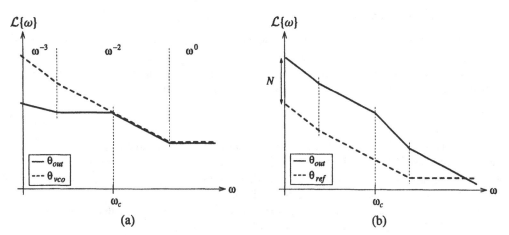

Figure 2.6: *The phase noise transfer functions in a PLL frequency synthesizer for (a) the VCO noise and (b) the reference noise.*

ω_c is the cross-over frequency of the PLL, i.e. the frequency at which the open loop gain is 0 dB. This frequency corresponds to the 3-dB cut-off frequency of the closed loop response of the PLL. For a constant loop filter the cross-over frequency is:

$$\omega_c = \frac{K_o}{N} \tag{2.18}$$

Looking at Eq. (2.16), the phase noise of the VCO is high-passed to the output of the PLL. High frequency noise passes unaltered, because the feedback gain of the loop is too low to suppress the noise at these frequencies. For lower frequencies, the feedback action of the loop kicks in and the noise is successfully suppressed. ω_c is the 3-dB point of this action. The VCO phase noise transfer is shown in Fig. 2.6 (a). The dashed line represents a typical VCO phase noise spectrum, which is discussed later in great detail (see Chapter 4). For now, it is sufficient to know that there are typically three noise regions; At high offset frequencies, a white phase noise floor can be distinguished. From there, the phase noise rises quadratically with the offset frequency (ω^{-2}). This noise is in fact white device noise that is frequency modulated around the carrier in the VCO circuit. The -20 dB/dec slope originates from the fact that frequency is the derivative of phase. Close to the carrier, frequency modulated 1/f noise determines the ω^{-3} region. The typical 1/f noise corner is not equal to the $1/f^3$ noise corner, but depends on non-linearities in the VCO amplifier (see Section 4.5.4). The resulting PLL output is shown in Fig. 2.6 (a) as the solid line.

The noise from the reference is low-passed to the PLL output with the same cross-over frequency, but within the PLL band the noise is amplified by the division factor, N. The resulting phase noise spectra are depicted in Fig. 2.6 (b). The reference noise exhibits the same three region as the VCO noise, but with a much smaller magnitude. This noise can corrupt the PLL output phase noise, since it is amplified by N for frequencies below ω_c. Noise from other building blocks in the PLL is also low-passed towards the output with a gain depending on their position in the loop.

2.3.1.3 rms Phase Error

Due to the filtering action in the PLL, the output spectrum of the stand-alone oscillator (Fig. 2.3) is shaped; In the loop bandwidth, the VCO noise is suppressed and the noise sources in the other loop components form an in-band phase noise floor, i.e. a flat region as in Fig. 2.6 (a). Combining Eq. (2.8) and Eq. (2.10), the phase noise formula becomes:

$$\mathcal{L}\{\omega_m\} = 20 \log \left(\frac{\theta_m(\omega_m)}{2} \right) \tag{2.19}$$

When assuming that the phase noise floor is flat in the band of interest, the in-band noise can be written as $\mathcal{L}\{f_m\} = 10 \log A_n$. The bandwidth of interest is the *noise bandwidth* of the synthesizer, f_{nbw} and is defined as the bandwidth of an equivalent phase noise spectrum with A_n for $f \leq f_{nbw}$ and zero for $f > f_{nbw}$ such that the total phase noise power is equal to that of the original spectrum. If $T_{cl}(s)$ is the closed loop transfer function for the noise that is low-passed to the output, the actual phase noise spectrum can be written using Eq. (2.19) as

$$\left(\frac{\theta_m(\omega_m)}{2} \right)^2 = |T_{cl}(j\omega_m)|^2 \cdot A_n \tag{2.20}$$

To find the overall rms phase error at the synthesizer output the rms value of $\theta_m(\omega_m)$ must be integrated over all offset frequencies. Since $\theta_m(\omega_m)$ is a sinusoidal signal, the rms phase error is given by $\theta_{rms}(\omega_m) = \theta_m(\omega_m)/\sqrt{2}$. The overall rms phase error is then

$$\Delta\Phi_{rms} = \int_0^\infty \frac{\theta_m^2(\omega_m)}{2} df_m \tag{2.21}$$

By integrating Eq. (2.20) and substituting the result of Eq. (2.21), the rms phase error becomes

$$\frac{\Delta\Phi_{rms}^2}{2} = A_n \left[\frac{1}{2\pi} \int_0^\infty |T_{cl}(j\omega_m)|^2 d\omega_m \right] \tag{2.22}$$

The integral in the above expression is the noise bandwidth of the PLL, i.e.

$$f_{nbw} = \frac{1}{2\pi} \int_0^\infty |T_{cl}(j\omega_m)|^2 d\omega_m \tag{2.23}$$

The noise bandwidth of a first-order synthesizer with a bandwidth f_c can be calculated to be $\frac{\pi}{2} f_c$. The noise bandwidth is 57% higher than the actual bandwidth to take into account all noise in the -20 dB/dec slope. The noise bandwidth of systems with a higher order the noise bandwidth is close to f_c.

Using Eq. (2.23), the rms phase error becomes

$$\Delta\Phi_{rms} = \sqrt{2 \cdot A_n \cdot f_{nbw}} \tag{2.24}$$

$$\cong \sqrt{2 \cdot A_n \cdot f_c} \tag{2.25}$$

The approximation in Eq. (2.25) holds for PLLs with higher order loop filters.

The rms phase error is related to the output signal-to-noise ratio (SNR_{PLL}) of the synthesizer [Best97]:

$$SNR_{PLL} = \frac{1}{2 \cdot \Delta\Phi_{rms}^2} \tag{2.26}$$

2.3.1.4 Spurious Suppression

Apart from distributed noise, the spectrum of an oscillator can also suffer from discrete spurious tones. If somewhere in the circuit, perturbations (through substrate or supply coupling or signal pick-up) at a fixed frequency generate an excess phase error $\theta_{sp}(t) = \theta_p \sin(\omega_m t)$, this phase error results in spurious tones at both sides of the carrier. Assuming again that $\theta_p \ll 1$, fulfilling the narrow band FM condition, the output of the VCO becomes:

$$V_{out}(t) \cong A \cdot \sin(\omega_0 t) + A \cdot \frac{\theta_p}{2} \cdot \left[\sin\left((\omega_0 + \omega_m)t\right) + \sin\left((\omega_0 - \omega_m)t\right) \right] \qquad (2.27)$$

When a pair of spurious tones with an amplitude A_{sp} show up in the output spectrum, the phase deviation is given by $\theta_p = 2 \cdot A_{sp}/A$. In other words the peak phase deviation depends on the ratio of the magnitudes of the spurious tones and the output signal and is expressed in dBc.

The spurious suppression is in first-order determined by the suppression of the loop filter. Since the suppression of higher order loops is usually sufficient, the spurious tones are mostly due to second-order effects, such as mismatches in the phase detector and parasitic injection of signals in the VCO due to substrate and power supply line coupling.

In communication systems, these spurious tones can act as pirate local oscillator signals in the reception path that downconvert unwanted signal onto the wanted signal. Similarly, in the transmitter these tones can upconvert unwanted signals and therefore they need to be within a pre-specified power mask.

2.3.2 Loop Dynamics

The derivations of the previous section were based on the steady-state operation of the PLL : a constant reference frequency is applied and the division modulus is fixed. However, the dynamic behavior of the loop at startup or when the reference frequency and/or the division modulus is changed is an equally important design parameter.

2.3.2.1 Tracking and Settling

First, the phase error $\theta_{err}(t) = \theta_{ref}(t) - \theta_{div}(t)$ for a specified input $\theta_{ref}(t)$ is studied, under the assumption that the loop is in lock, and that the phase error is sufficiently small to justify linear calculus. Most important is the steady-state phase error $\theta_{err,ss}$, i.e. the value that $\theta_{err}(t)$ assumes after all the transients have died away. It is calculated using the final value theorem of the Laplace transformation which states that

$$\lim_{t \to \infty} \theta(t) = \lim_{s \to 0} s\theta(s) \qquad (2.28)$$

If the reference input is a phase step of magnitude $\Delta\theta$, the resulting steady-state phase error for a first-order loop is

$$\theta_{err,ss} = \lim_{s \to 0} \left[s \cdot \theta_{ref}(s) \cdot \frac{1}{1 + GH(s)} \right] = \lim_{s \to 0} \left[s \cdot \frac{\Delta\theta}{s} \cdot \frac{1}{1 + K_o/(Ns)} \right] = 0 \qquad (2.29)$$

So the first-order PLL reduces any phase error to zero. If the input frequency changes with a step of size $\Delta\omega$, the resulting steady-state phase error is

$$\theta_{err,ss} = \lim_{s \to 0} \left[s \cdot \frac{\Delta\omega}{s^2} \cdot \frac{1}{1 + K_o/(Ns)} \right] = \frac{\Delta\omega \cdot N}{K_o} = \frac{\Delta\omega}{\omega_c} \tag{2.30}$$

A frequency step happens when suddenly the division modulus in the synthesizer is changed in order to obtain a new output frequency. To keep the final error small, the forward gain or the bandwidth must be as high as possible.

Apart from the steady-state behavior, it is also necessary to determine the transient phase error caused by particular inputs. Again, the most important situation in a frequency synthesizer is the one where the division modulus N is changed, which is equivalent to a change in input frequency $\Delta\omega$. The time the first-order loop needs to settle to the wanted output frequency is:

$$\theta_{err}(s) = \frac{1}{1 + GH(s)} \cdot \theta_{ref}(s) = \frac{s}{s + \omega_c} \cdot \frac{\Delta\omega}{s^2} = \frac{\Delta\omega}{s \cdot (s + \omega_c)} \tag{2.31}$$

$$\theta_{err}(t) = \frac{\Delta\omega}{\omega_c} \cdot \left(1 - e^{-\omega_c t} \right) \tag{2.32}$$

So the final frequency is obtained after an exponential behavior with time constant $1/\omega_c$. With these equations, the time T_ε to settle the loop to the new output frequency within a specified accuracy ε is found to be

$$T_\varepsilon = -\frac{\ln \varepsilon}{\omega_c} \tag{2.33}$$

with $\varepsilon = \Delta\omega_{acc}/\Delta\omega$ and $\Delta\omega_{acc}$ the frequency accuracy.

2.3.2.2 Acquisition

In the previous derivations, it is assumed that the loop is already in lock. But every loop starts in an unlocked condition at power-up, and must be able to achieve lock, either by its own natural actions or with the help of auxiliary circuits. This process is called acquisition. Since acquisition is inherently a non-linear process, analytical calculation goes beyond the scope of this book. We will limit ourselves to some descriptive analysis. More information can be found in [Best97].

If the initial VCO period multiplied by N is sufficiently close to the period of the reference frequency, the PLL will lock up with just a phase transient; No cycle slipping occurs prior to lock. The frequency range over which this is possible is called the *lock-in range* $\Delta\omega_L$.

The *pull-in range* $\Delta\omega_P$ is defined as the range of frequencies over which the PLL will acquire lock after slipping cycles for a while. It is always larger than the lock-in range.

Finally, the frequency range over which the loop will hold lock is defined as the *hold-in range*. A PLL is conditionally stable within this range. It is larger than both other defined ranges. For example, in (2.30) the linear approximation of phase error due to a frequency step is shown to be $\theta_{err,ss} \propto \Delta\omega/K_o$. However, a real phase detector does not have an infinite linear range, as will be discussed in section 2.4.1. The PD-dependent hold-in range therefore lies between $\Delta\omega_H = \pm K_o$ (for a sinusoidal PD) and $\Delta\omega_H = \infty$ (for a phase-frequency detector). These definitions are only valid as long as the limit is set by the PD and not by some other non-linearity, such as clipping in an operational amplifier or of the VCO control signal.

2.3.3 Other Specifications

- Tuning range: A wireless transceiver uses predefined frequency bands to transmit and receive, which must be accommodated by the frequency synthesizer. The tuning range, i.e. the frequency range that can be synthesized, is mainly determined by the tuning range of the VCO.

- Integratability: is the measure of how well the synthesizer can be integrated. The typical integratability bottlenecks in high-quality synthesizer design are the high Q inductor and the loop filter capacitance. In order for the PLL to occupy a moderate area, the integrated capacitance of the loop filter must be minimized while maintaining the required dynamic behavior and phase noise performance.

- Power consumption: The power consumption must of course be as low as possible for portable telecommunication systems.

- Frequency resolution: The frequency resolution specification is set by the required channel spacing of the intended application. For integer-N PLL the frequency resolution is determined by the reference frequency, and sets the loop bandwidth for spurious suppression.

2.4 Introduction to PLL building blocks

2.4.1 The Phase Detector

Basically, three flavors of phase detection exist; The analog phase detector of multiplier performs a mixing operation on its input signals and the resulting DC output is a measure for the phase error. Digital phase detectors are implemented using EXOR gates or in a sequential way with flipflops that trigger on the zero-crossings of their inputs. The final category is the phase-frequency detector, which is digital and sequential and provides apart from phase detection also frequency detection to aid acquisition.

2.4.1.1 Analog Phase Detectors

Analog phase detector operate typically on sinusoidal input signals. The two inputs $A_1 \cdot \sin(\omega_1 + \theta_1)$ and $A_2 \cdot \sin(\omega_2 + \theta_2)$ are multiplied or mixed resulting in an output voltage v_{pd} (see Eq. (2.34)).

$$v_{pd} = A_{pd} \cdot \left(\sin[(\omega_1 - \omega_2)t + \theta_1 - \theta_2] + \sin[(\omega_1 + \omega_2)t + \theta_1 + \theta_2] \right) \qquad (2.34)$$

When both frequencies are the same, the DC component of the output equals $A_{pd} \cdot \sin(\theta_1 - \theta_2)$. Unwanted signals are present at twice the reference frequency (the sum-component) and at the reference frequency itself with a magnitude depending on the LO-to-IF and RF-to-IF isolation of the mixer.

The analog phase detector is especially useful in applications where the reference frequency is too high for other circuits and where the loop bandwidth is sufficiently narrow to effectively suppress the unwanted signals.

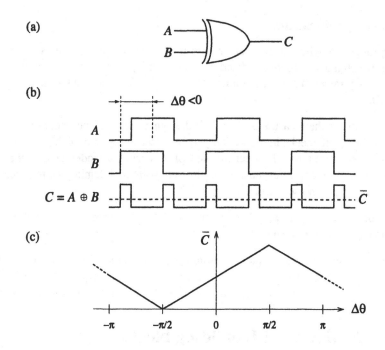

Figure 2.7: *The EXOR phase detector (a), operation (b) and transfer characteristic (c).*

2.4.1.2 The EXOR Phase Detector

An EXOR gate can be used as a phase detector as shown in Fig. 2.7 (a). The operation is illustrated for the case that the phase difference $\Delta\theta$ between the inputs A and B is negative. Lock is obtained when both input signals are $-90°$ shifted. The average value of the output $\bar{C}$ is a measure for the phase difference. The corresponding transfer characteristic is plotted in Fig. 2.7 (c). Around zero phase difference a linear operating range is situated that stretches from $-\pi/2$ to $\pi/2$. The output of the EXOR gate contains no energy at the reference frequency, but the component at twice the reference frequency reaches its maximum amplitude, i.e. $4/\pi$ times the peak-to-peak output, at zero phase error. Unfortunately this is the operating point of the PD. Another problems arises when both inputs are asymmetrical; In that case the output signal gets clipped around $-\pi/2$ and $\pi/2$, reducing the loop gain of the PLL and thus the locking capabilities.

2.4.1.3 Flipflop Phase Detectors

A phase detector using a JK-flipflop is shown in Fig. 2.8. The inputs A and B set and reset the output of this sequential circuit. The operation is again plotted for a negative phase difference. The average value of the output, $\bar{C}$ has a sawtooth characteristic. The linear operating range of this phase detector is a full reference period and is centered around $\pm\pi$ radians. When the

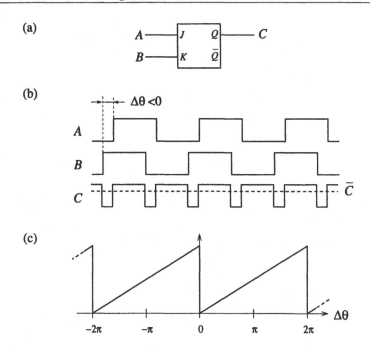

Figure 2.8: *The JK circuit (a), operation (b) and transfer characteristic (c).*

loop is in lock, the output of the phase detector has a component at the reference frequency with a magnitude of $4/\pi$ times the peak-to-peak range of the PD. Compared to the EXOR PD, the linear operating range is doubled, but the flipflop is sensitive to reference spurs. Of course, the same phase characteristic can be realized with a SR flipflop.

2.4.1.4 The Phase-Frequency Detector

The phase-frequency detector is also a sequential phase detector, but it has an extra feature to it; The PFD has a memory function, such that it can act not only as a phase detector but also as a frequency detector. A typical circuit implementation of a PFD is shown in Fig. 2.9 (a). The PFD has two outputs, Up and Dn control a charge pump that succeeds the PFD. When the Up signal is high, the charge pump dumps current in the loop filter causing a rising voltage at the VCO input, such that the frequency increases. The Dn signal has the opposite effect and draws current out of the loop filter impedance causing a control voltage drop. A third state is when none of the signals is active; The output current is then zero and the output is a high-impedance node.

The operation of the PFD is illustrated in Fig. 2.9 (b). This PFD circuit triggers on the falling edges of the input signals. In the initial phase, the Div input lags the Ref input and an Up pulse is generated to increase the VCO frequency. At the next sample moment, both edges occurs simultaneously and no current is dumped in the loop filter impedance. It is clear that the frequency of the Div signal is higher than that of the reference input and at the sampling

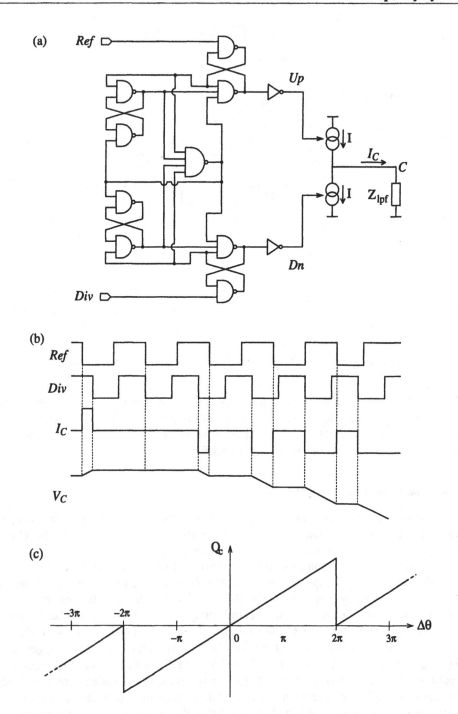

Figure 2.9: *A typical phase-frequency detector implementation (a), operation (b) and transfer characteristic (c).*

moments that follow a Dn pulse with increasing width is generated. As a result, more and more current is drawn out of the loop filter impedance and the control voltage V_C to the VCO decreases (see Fig. 2.9 (b)), lowering its output frequency. The total charge Q_c dumped in the loop filter for a certain phase difference is plotted in (c), revealing the phase transfer characteristic. The linear operating range is now 4π radians, which is twice as large as the for the flipflop PFD. What's more, at lock, ideally no control pulses occur, such that the output contains no spurs. Fig. 2.9 (c) clearly shows the frequency detection capability of the PFD. When the phase error is larger than $\pm 2\pi$, the PFD output keeps sending the VCO in the right direction. This means that the hold range of a PLL with a PFD is limited by the PFD itself but by the other building blocks.

A major drawback of this kind of circuits is the possible occurrence of a dead zone near zero phase error. This means that the loop is not able to react to small phase changes. As a result, when the loop is locked, phase noise very close to the carrier can pass unaltered since the loop gain is zero for very small phase deviations. Remedies to counter the dead zone problem are discussed in Chapter 6. Another problem that occurs in real implementations is parasitic charge injection in the loop filter impedance, at the sampling moments, even when the loop is locked; Therefore some energy at the reference frequency might show up at the synthesizer output.

2.4.2 The Loop Filter

The design of the loop filter determines most of the specifications of the PLL. Extra poles and zeros in the loop transfer function influence the noise and dynamic performance of the loop. In the previous section, the first-order loop $G_{lpf}(s) = K_{lpf}$ is already discussed. From Eq. (2.18), it is clear that good tracking and narrow loop bandwidth are incompatible for the first-order PLL; Therefore it is not often used. In what follows the first-order loop is extended to higher orders and types to improve the PLL performance. The highest power of s in the denominator of the transfer function determines the PLL *order*. The *type* of the loop is the number of perfect integrators in the loop. Every PLL is at least type-I due to the perfect integration in the VCO.

2.4.2.1 Second-Order PLLs

The most simple second-order PLL is realized by placing a pole in the loop filter with a frequency higher than ω_c. When the pole frequency is sufficiently low it provides additional suppression of spurious tones at the reference frequency generated by the phase detector. The loop filter transfer function is then $G_{lpf}(s) = (1 + s\tau)^{-1}$ with $\tau = RC = 1/\omega_p$. The closed loop transfer function for θ_{ref} is

$$\frac{\theta_{out}(s)}{\theta_{ref}(s)} = \frac{G(s)}{1 + GH(s)} = \frac{\omega_p K_o}{s^2 + \omega_p S + \omega_p K_o/N} \tag{2.35}$$

In the standard notation for second-order systems, the closed loop transfer function becomes

$$\frac{\theta_{out}(s)}{\theta_{ref}(s)} = \frac{\omega_p K_o}{s^2 + 2\zeta \omega_n S + \omega_n^2} \tag{2.36}$$

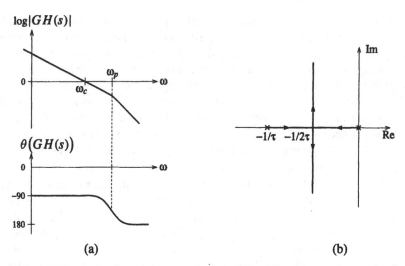

Figure 2.10: *(a) The bode plot of the open loop gain and (b) the root-locus plot for a second-order, type-I PLL.*

with a natural frequency and a damping factor given by

$$\omega_n = \sqrt{\omega_p K_o / N} \qquad \zeta = \frac{1}{2} \cdot \sqrt{\omega_p N / K_o} \qquad (2.37)$$

The bode plot of the open loop gain of the second-order PLL is given in Fig. 2.10 (a). Since only one integration is present in the loop, it is a type-I loop. The second-order PLL can never become unstable, since the maximum phase shift is 180° for any finite frequency.

To gain more insight in the behavior of the PLL, the closed loop poles and zeros are plotted for the loop gain going from 0 to ∞, i.e. the *root-locus* (see Fig. 2.10 (b)). The plot starts at the open-loop poles (zero loop gain) and terminates on the open-loop zeros (infinite gain). Initially, the poles move toward each other on the negative real axis. When they meet halfway, they become a complex conjugate pair and move towards infinity along a vertical line at $s = -1/2\tau$. The damping becomes very small as the gain increases.

To evaluate the dynamics of the loop, the steady-state error after a phase and frequency step is investigated using Eq. (2.29) and Eq. (2.30). Since the loop is a type-I system, $\theta_{err,ss,\Delta\theta} = 0$ and $\theta_{err,ss,\Delta\omega} = N \cdot \Delta\omega / K_o$.

The simple second-order PLL provides only one design parameter: ω_p. Therefore, if a large DC gain is needed for a small steady-state error for a frequency step and a small bandwidth, the system becomes seriously underdamped, causing ringing in the transient response. The solution for this problem is *lag-lead* compensation. The lag-lead filter has two independent time constants, such that ω_n and ζ can be chosen independently. Furthermore, high DC gain and good tracking are possible. Two typical compensated second-order PLL loop filter implementations exist: a passive and an active filter. Both filter implementation are shown in Fig. 2.11 (a) and

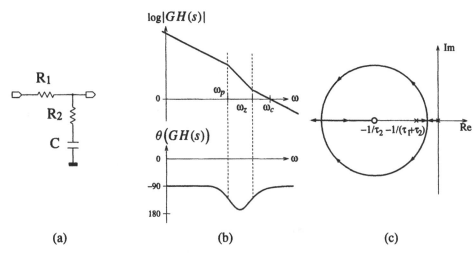

(a)   (b)   (c)

Figure 2.11: *(a) A passive lag-lead filter with the corresponding (b) bode plot of the open loop gain and (c) the root-locus plot for a second-order, type-I PLL.*

Fig. 2.12 (b). The filter transfer function of the passive filter is

$$G_{lpf}(s) = \frac{1 + s\tau_2}{1 + s(\tau_1 + \tau_2)} \tag{2.38}$$

with $\tau_1 = R_1C$ and $\tau_2 = R_2C$. In the standard notation , the closed loop transfer function for high loop gains becomes

$$\frac{\theta_{out}(s)}{\theta_{ref}(s)} = \frac{2\zeta\omega_n s + \omega_n^2}{s^2 + 2\zeta\omega_n S + \omega_n^2} \tag{2.39}$$

with a natural frequency and a damping factor given by

$$\omega_n = \sqrt{K_o/(\tau_1 + \tau_2)} \qquad \zeta = \frac{\omega_n}{2} \cdot (\tau_2 + 1/K_o) \tag{2.40}$$

The bode plot of the open loop gain is plotted in Fig. 2.11 (b) with $\omega_p = 1/(\tau_1 + \tau_2)$ and $\omega_z = 1/\tau_2$. The additional time constant adds a high frequency zero, which increases the phase margin and therefore improves the transient response. The root-locus is shown in (c); Again with increasing loop gain, both closed loop poles move towards each other and form a complex pair. Because of the zero, the complex portion of the root-locus is a circle centered around $1/(\tau_2)$. For small gains the damping is small, but it increases with higher gains. For very high gains, the locus returns to the real axis and the loop becomes overdamped. The second-order PLL with a passive lag-lead filter is still a type-I loop. The steady-state errors remain the same as before.

The active implementation of the loop filter is shown in Fig. 2.12. This time the lag-pole approaches zero (due to the high gain of the opamp), adding an extra integration in the loop. This type of filter is also called a *PI (proportional-plus-integral)* filter. The bode plot and root-locus are shown in Fig. 2.12 (b) and (c) and are similar to those of the passive lag-lead filter. The

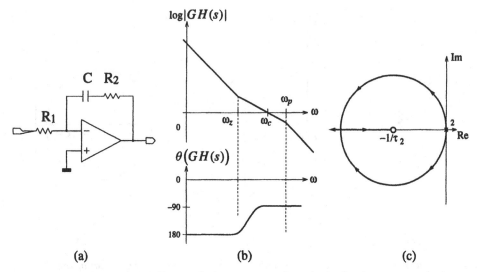

Figure 2.12: *(a) An active PI filter with the corresponding (b) bode plot of the open loop gain and (c) the root-locus plot for a second-order, type-II PLL.*

filter transfer function of the active PI filter is

$$G_{lpf}(s) = -\frac{1 + s\tau_2}{s\tau_1} \tag{2.41}$$

with $\tau_1 = R_1 C$ and $\tau_2 = R_2 C$. In the standard notation , the closed loop transfer function for high loop gains is the same as in Eq. (2.39) with a natural frequency and a damping factor given by

$$\omega_n = \sqrt{K_o/(\tau_1 + \tau_2)} \qquad \zeta = \frac{\tau_2 \omega_n}{2} \tag{2.42}$$

Since the type of the loop has been changed by adding the active element in the filter, the tracking behavior of the loop is also altered. The steady-state error after a phase change (Eq. (2.29)) is still zero. But due to the extra integration, the steady-state error after a frequency step Eq. (2.30) is also zero. This means that the loop can track phase and frequency steps with zero error. As a result, the hold range $\Delta\omega_H$ is no longer restricted by the loop gain, but by the frequency range of the oscillator.

To evaluate the noise performance of the second-order PLL, the noise bandwidth of the loop [Rhod83] is of importance:

$$f_{nbw} = \frac{\omega_n}{2}\left(\zeta + \frac{1}{4\zeta}\right) \tag{2.43}$$

The noise bandwidth is minimal for $\zeta = 0.5$. However, for optimal transient response and a flat closed loop transfer function, the damping factor is $\zeta = 1/\sqrt{2} = 0.707$, in which case the closed loop poles lie on radial lines $\pm 45°$ from the real axis. Since the function $f_{nbw}(\zeta)$ is flat around $\zeta = 0.5$, a damping factor of around 0.7 still is the optimal solution.

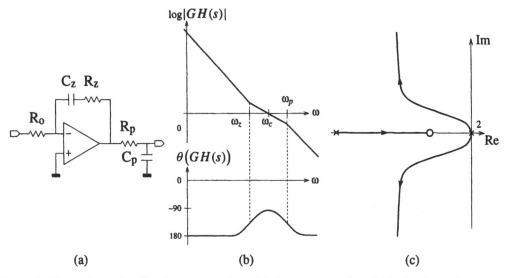

Figure 2.13: *(a) An active filter implementation with the corresponding (b) bode plot and (c) an example of a root locus of the open loop gain for a third-order, type-II PLL.*

2.4.2.2 Third-Order PLLs

To further increase the number of design parameters, a third pole can added in the loop filter of the type-II, second-order PLL (see Fig. 2.13 (a)). An active implementation is again necessary to make the loop a type-II loop. The extra pole again increases the spurious suppression of the loop. Since this pole gives an extra $-90°$ phase shift, it must be placed at higher frequencies than the zero, to maintain a sufficient phase margin. The zero is shifted in front of the cross-over frequency. In Fig. 2.13 (b), the bode plot of the open loop gain is plotted, showing the extra high frequency suppression and pole-zero positioning for high phase margin. The pole and zero positions are given by $\omega_z = 1/(R_z C_z)$ and $\omega_p = 1/(R_p C_p)$. In Fig. 2.13 (c), an example of a root locus plot of the given implementation of a third-order, type-II PLL. The zero and high frequency pole are placed to provide a phase margin higher than $60°$. The two poles at the origin form a complex conjugate pair for higher loop gains. Note that in contrast to a second-order loop, the third-order loop can become unstable, if not well implemented. The steady-state errors are the same as for the second-order, type-II loop, but the noise performance is enhanced.

2.4.3 The Oscillator

The oscillator in a PLL operates at the highest frequency and usually determines the out-of-band noise of the frequency synthesizer (see Eq. (2.13)). The most important specifications of the oscillator are the phase noise, the tuning range, the power consumption, frequency pushing/pulling and the cost. Frequency pushing is the dependency of the output frequency on the power supply, while frequency pulling is the dependency on the output load impedance.

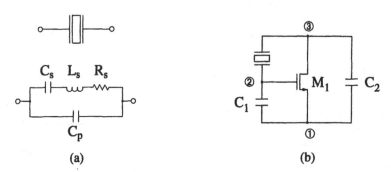

Figure 2.14: *(a) The symbol and equivalent circuit of a crystal. (b) The basic three-point crystal oscillator circuit without biasing circuits; Depending on the position of the AC ground, the circuit is a ① Pierce, ② Colpitts or ③ Santos oscillator.*

2.4.3.1 The Xtal Oscillator

The most stable oscillator is the crystal oscillator, which exploits the clean electro-mechanical resonance of a piezoelectric crystal (Xtal). The symbol and equivalent circuit of a crystal is shown in Fig. 2.14 (a). Typical values for the equivalent model are: $C_p = 6\text{pF}$, $C_s = 30\text{fF}$, $L_s = 8.4\text{mH}$ and $R_s \approx 5.3\Omega$ with a resonance frequency of 10 MHz and a quality (Q) of around 10^5. From the same figure can be derived that a crystal has two resonance modes. At series resonance, i.e. when L_s cancels C_s, the impedance of the crystal is close to zero. At parallel resonance, i.e. when the impedance of the series branch is equal to that of C_p, the impedance of the crystal becomes infinite. The corresponding resonance frequencies ω_s and ω_p are:

$$\omega_s^2 = \frac{1}{L_s C_s} \qquad \omega_p^2 = \frac{1}{L_s} \cdot \left(\frac{1}{C_s} + \frac{1}{C_p} \right) \tag{2.44}$$

Since C_p is much larger than C_s, the parallel and series resonance frequencies are very close together. This is the reason why crystal oscillators exhibit very low phase noise; Due to the high capacitance ratio, large voltages are generated internally in the crystal, leading to an excellent signal-to-noise ratio.

The basic three-point oscillator circuit to build an integrated crystal oscillator with only the crystal as an external component is shown in Fig. 2.14 (b) without biasing and extra circuitry. Depending on the position of the ground, the circuit is a ① Pierce [Vit88], ② Colpitts [Hua00] or ③ Santos [San84] oscillator. The design of all three circuits is based on the general theory developed in [Vit88] that allows accurate linear and non-linear analysis of the crystal oscillator. The Pierce oscillator requires two output pins for the Xtal, while the crystal in the Colpitts and Santos oscillator is connected by one pin to the circuit and one pin to the common ground. Therefore, the latter types are preferred for clock generation in digital ICs for minimizing the number of pins. The frequency stability of the Pierce oscillator is however superior [Vit88]. Typical resonance frequencies of crystals are in the range of some tens of MHz. They are typically used as a reference source in PLL synthesizers. To compensate drift of the crystal properties over temperature, Temperature-Controlled Crystal Oscillators (TCXO) exist in which the crystal is kept to a constant temperature by a mini-heater. By placing a voltage controlled capacitor in parallel with C_p, a voltage controlled XO (VCXO) can be built.

2.4.3.2 The Relaxation Oscillator

The relaxation oscillator is well suited for silicon integration and high speed operation. The basic circuit consists of a timing capacitor that is charged and discharged. A triangle wave is obtained across the capacitor and square waves at the output. By changing the biasing currents, very wide tuning ranges can be achieved. Different implementation in different technologies (also CMOS [Raz96]) exist, but the operation principle remains the same.

Theory on the phase noise in relaxation oscillators can be found in [Abid83, Snee90]. The theory predicts that the phase noise is proportional to the equivalent noise generating resistor. To lower the phase noise, the resistance must be lowered, but this implies a larger power consumption. Typical specs for relaxation oscillators are : for 100 MHz, a phase noise of -118 dBc/Hz at 1MHz [Snee90] or equivalently for 920 MHz, a phase noise of -102 dBc/Hz at 1MHz [Raz96].

2.4.3.3 The Ring Oscillator

The ring oscillator is one of the most popular oscillators for integrated PLLs and clock recovery circuits since it is less complex and easy to integrate. The periodic output signal is generated by a ring of inverters. Three or more inverters can be used to build the ring, but the number must be odd to guarantee oscillation. The oscillation output period is $2n \cdot T_d$ with n the number of inverters and T_d the delay in one inverter, also called a delay cell. Tuning is achieved by varying the current to charge the output capacitance of the inverters; The resulting tuning range is large.

The constraint of an odd number of inverter poses a limit on the highest output frequency, since at least three inverters are necessary to realize oscillation. Differential inverters (i.e. differential pairs) with cross-coupled connections allow the use of only two inverters in the ring. The higher circuit complexity limits the speed enhancement of these topologies. To enhance the high frequency operation and to increase the tuning range, a topology with dual-delay paths is developed in [Park99]; A negative skewed delay path decreases the unit delay in the ring and by combining the negative skewed delay and the normal delay path, the frequency range is increased. A distinct advantage of the even-stage differential ring oscillator is the intrinsic availability of quadrature outputs [Abid94].

Ring oscillators are very well suited for CMOS integration and are capable of GHz operation [Raz96, Haji99]. Unfortunately, the phase noise of ring oscillators is inferior, impeding its use in high-quality communication systems. The lack of noise filtering (as in LC-tank oscillators) and the switching action in the delay cells cause the intrinsically high phase noise. Again, the phase noise is inversely proportional to the power consumption. More elaborate discussions on ring oscillator phase noise are given in [Haji99, Wei94, Raz96]. Typical phase noise values are -95 dBc/Hz at 1 MHz offset and 1.8 GHz output [Haji99]. State-of-the-art designs report a phase noise as low as -117 dBc/Hz at 600 kHz offset for an output frequency of 900 MHz, a tuning range of 50% and a power consumption of 30 mW [Park99].

2.4.3.4 The LC Oscillator

Both relaxation oscillators and ring oscillators are very well suited for silicon integration. They offer high frequency operation and a high tuning range. However, both oscillator types suffer from bad phase noise performance. To improve the phase noise to levels needed in high-quality wireless communication systems, the power consumption becomes unacceptably high. The best way to achieve monolithic integration of the oscillator is the LC-oscillator; Its output frequency is determined by the resonance of an inductor and a capacitor. Since both elements constitute a passive filter, the phase noise is expected to be low. A phase noise analysis is given in Chapter 4, showing that low phase noise and low power are compatible for LC-oscillators. Typically, the phase noise of LC-oscillators is 20 dB better than that of relaxation or ring oscillator. Very high output frequencies are achievable, since all parasitic capacitances can be tuned out by the inductor. Therefore, this type of oscillator is the subject of this book (see Chapter 4).

The bottleneck of fully integrated LC-tank oscillators is the integrated inductor. Most designs employ off-chip, high Q components [Soy89] or external transmission lines on the PCB, acting as an inductor [Pfaf02]. Since both solution require extra components and processing, lots of effort has been put in achieving integrated solutions; Active inductors can be implemented on-chip [Wang90], but they are usually too noisy and power-hungry for high-quality designs. Also bondwire inductors are feasible since they are part of the packaging process [Cran95a, Svel00], but they lack robustness and reproducibility. The most interesting solution are integrated planar inductors, the integration of which is discussed in Chapter 4 in great detail.

2.4.3.5 Other Oscillator Types

Another type of integratable oscillator circuit is the OTA-C oscillator; Oscillation is realized by positive feedback around an active element (typically an OTA) using only capacitors and resistors. The most classical example is the Wien-bridge oscillator [Sen89].

More exotic, high frequency oscillator types, which do not qualify for silicon integration exist such as Dielectric Resonance Oscillators (DRO), Yttrium Iron Garnet (YIG), ...

2.4.4 The Frequency Divider

The frequency divider is together with the oscillator, the only part of the PLL frequency synthesizer that operates at high frequency. It converts the high oscillator output frequency to lower frequency which can be compared to the clean Xtal reference source.

A frequency divider is basically a counter, whose output state changes after it has counted a predefined number of input periods. The simplest divider is a flipflop, which performs a divide-by-2 operation on the input frequency. The functional input of the flipflop for frequency division is the clock input. Two types of frequency divider circuits exist: the asynchronous divider and the synchronous divider; The asynchronous divider consists of a cascade of flipflops, whose

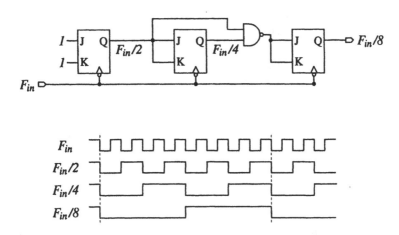

Figure 2.15: *A synchronous divide-by-8 frequency divider with JK flipflops.*

outputs are fed to the clock input of the next divider. The input signal to this divider triggers the first flipflop, which then triggers the next one, ... The synchronous divider is also a cascade of flipflops but this time they are all triggered by the input signal. Since in the asynchronous counter, the signal has to ripple through the total flipflop chain, the output changes with a delay with respect to the input signal.

An implementation example of a synchronous counter with JK flipflops is shown in Fig. 2.15. All flipflops change state simultaneously on the edge of the input clock, in contrast to the asynchronous divider. To avoid race problems creating glitches which on their turn cause spurious tones, the input timing to the NAND gates in such a counter is critical.

2.4.4.1 Programmable Dividers

In most synthesizer applications, the output frequency and thus the division modulus needs to be programmable. For low frequency operation, programmable division is realized using programmable counters, while high-frequency operation is obtained by employing prescalers.

A programmable counter is programmed by a preset number P and starts counting input pulses until the end number is reached and an overflow signal is generated. The counter is then reset to the preset number and starts counting again. The resulting division ratio is $2^n - P$, with n the number of bits of the counter. Another possible implementation is that the counter counts till P is reached and is reset. The division ratio then equals P.

If the input frequency to the divider is too high for digital programmable counters, *prescalers* are used. A prescaler divides by a fixed number, omitting the delay problems of the programmable counter. In fact, the prescaler pre-scales the input frequency for the subsequent programmable counter stages. The main disadvantage of fixed prescaler is that for a given frequency resolution, the reference frequency needs to be lowered. For a prescaler division by N_p the smallest possible frequency step of the synthesizer becomes $N_p \cdot f_{ref}$. This means that the

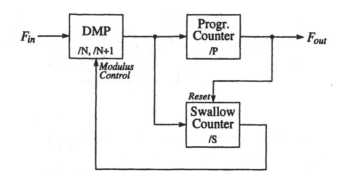

Figure 2.16: *A full frequency divider with a dual-modulus prescaler and two counters.*

reference frequency, and thus the loop bandwidth needs to be a factor N_p lower, which is most undesirable in integrated synthesizer applications.

The solution to the frequency resolution problem is the dual-modulus prescaler (DMP). The dual-modulus prescaler is able to divide by N_p and $N_p + 1$ with some additional logic. The additional logic adds some delay in the circuit, reducing its operating speed. DMP implementations are more elaborately discussed in Chapter 3. By combining the dual-modulus prescaler with the proper programmable counters, a programmable high-frequency divider can be constructed (see Fig. 2.16). The full divider consists of a DMP, a programmable counter P and a swallow counter S. The DMP divides by $N_p + 1$ and the S-counter counts the output pulses, until a number S is reached. It then changes the DMP modulus control bit, resetting the prescaler division to N_p. The P-counter also counts the DMP output pulses, until a number P is reached. It then resets both the S- as the P-counter, and the division process is restarted. This means that during one output period of the full divider, the DMP has divided S times by $N_p + 1$ and $P - S$ times by N_p, such that the overall division number becomes:

$$N = (N_p + 1) \cdot S + N_p \cdot (P - S) = P \cdot N_p + S \qquad (2.45)$$

If S is variable between 0 and $N_p - 1$, the complete range of division numbers can be realized. For proper reset by the P-counter, P must be larger than the largest value of S, i.e. $P \geq N_p$. For a given minimum synthesizable frequency, the prescaler division number is limited, since the smallest obtainable division number is $N_{min} = N_p^2$.

2.5   Advanced PLL Frequency Synthesizers

Since the smallest frequency step of a conventional phase-locked loop is f_{ref} (see Eq. (2.1)), the loop bandwidth of the PLL needs to be typically smaller than one tenth of f_{ref} for stability and reference spurious suppression, limiting the dynamic performance of the loop. In an attempt to combine the best of both worlds, other synthesizer varieties have been developed, which are basically combinations of/or variations on the three basic synthesizer flavors: table-look-up synthesizers, direct digital synthesizers and indirect synthesizers.

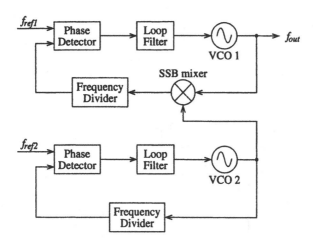

Figure 2.17: *A dual-loop PLL frequency synthesizer with suppressed mixing sidebands.*

2.5.1   Combining Frequency Synthesizers

A first example is Direct Digital Frequency Synthesis (DDFS) [Abid94]. In DDFS, a low frequency signal is generated by a table-look-up synthesizer, which is than upconverted to the desired high frequency output by a fixed frequency PLL synthesizer. Since the output of the PLL is
fixed, the loop bandwidth can be optimized towards noise. This configuration enables high output frequencies and very fast frequency hopping. The output frequency range is however rather
low and the spectral purity rather poor due to the DAC. This can be remedied by implementing a
high-speed, high-accuracy DAC, resulting in a very power-hungry solution.

A second example a dual-loop indirect synthesizer, which combines two phase-locked loop
synthesizers. The reference frequencies and loop bandwidths of both loops can be chosen separately, optimizing them to the best overall result. The major drawback of this configuration is the
fact that the two output frequencies must be mixed in a single-sideband (SSB) mixer, which requires accurate quadrature phases, low harmonic distortion and well-matched mixers [Raz97b].
To reduce the problems of the single-sideband mixing, the mixer can be placed inside the loop
as in Fig. 2.17. The frequency of the sidebands originating from mismatches and harmonics in
the mixing process is such that the sidebands are suppressed by the loop filter of the first PLL
[Raz97b]. The integration of this type of synthesizer can become quite bulky and power hungry,
due to the two loops that need to be implemented and due to the quadrature phase generation for
the SSB mixing.

Other possible synthesizer combinations are those in which the reference input to a PLL
synthesizer is another synthesizer. If a programmable reference input is synthesized by a table-
look-up synthesizer, a high-Q bandpass filter is necessary at the output of the table-look-up
synthesizer to remove spurious tones. All tones are amplified by the division modulus N in
the PLL and therefore the loop bandwidth of the PLL must be rather low for spurious suppression. This configuration is highly complex, hard to integrate due to the band-pass filter and

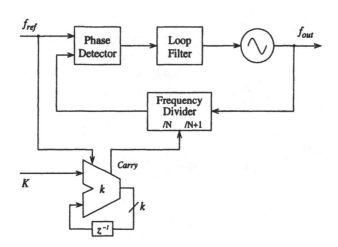

Figure 2.18: *Fractional-N frequency synthesizer with digital accumulator.*

power-hungry because of the high-speed, high-accuracy DAC for fine frequency resolution and sufficient frequency range. Another possibility is using a PLL as the reference synthesizer; The reference loop performs channel selection and has a small loop bandwidth, while the actual PLL has a large loop bandwidth [Vauc01]. The problem with this architecture is that noise from the reference synthesizer must be very low for the overall in-band noise and that the noise from the output VCO must be low for high offset noise. For integration, this results in a complex and power-hungry solution.

2.5.2 Fractional-N Frequency Synthesizers

In fractional-N frequency synthesizers, *fractional* multiples of the reference frequency can be synthesized, allowing a higher reference frequency for a given frequency resolution. The basic operation can be explained using the single digital accumulator (see Fig. 2.18); When a digital word K is applied to the k-bit digital accumulator, a carry bit is produced on average every $K/2^k$ cycles of the reference frequency f_{ref}. This carry bit controls the divider modulus. A dual-modulus prescaler will divide K cycles by $N+1$ and $2^k - K$ cycles by N, resulting in an average division number:

$$N_{frac} = \frac{\left(2^k - K\right) \cdot N + K \cdot (N+1)}{2^k} = N + \frac{K}{2^k} \qquad (2.46)$$

The generation of the carry bit and thus the modulus switching is a periodic event, creating large spurs in the PLL output spectrum. To alleviate these spurs, the phase information in the accumulator can be used to compensate the phase error at the PD output, using a DAC. This technique relies on analog matching of currents which must compensate, making it sensitive to temperature and process variations. Better is to replace the accumulator by an all-digital, higher-order $\Delta\Sigma$ modulator. The resulting $\Delta\Sigma$ bitstream, that controls the divider number, consists of a

DC-value that corresponds to the required fractional division number and of shaped quantization noise. Higher-order $\Delta\Sigma$ modulators remove all the quantization noise around DC and multiples of f_{ref} and provide good randomization of spurious energy. The residual quantization noise is then removed by the averaging action of the loop filter.

The main advantage of fractional-N synthesis is the decoupling of the reference frequency and the loop bandwidth. The loop bandwidth can be increased, without deteriorating the spectral purity. Therefore, the PLL dynamics are accelerated and the total amount of required capacitance in the loop filter can be decreased, such that single chip integration of the frequency synthesizer becomes feasible. The fractional-N synthesizer employs the same, straightforward architecture as the integer-N synthesizer, with additional digital logic to properly control the division moduli. Frequency resolution is traded against digital complexity. Since CMOS technology is the natural biotope of digital circuitry, the fractional-N synthesizer is an excellent option for full CMOS frequency synthesizer integration.

2.6 Frequency Synthesis for the DCS-1800 System

2.6.1 A Fully Integrated DCS-1800 Transceiver

From the previous section it became clear that to build a fully integrated transceiver for the DCS-1800 communication system (see Section 2.6.2), the fractional-N frequency synthesizer with $\Delta\Sigma$ control offers the most integrated solution and the best path to full CMOS integration:

- The fractional-N PLL synthesizer offers a potentially higher bandwidth by increasing the reference frequency, enabling the full integration of the loop filter, without blowing up the IC area.

- In a $\Delta\Sigma$ fractional-N synthesizer frequency resolution is traded against digital complexity, which is the natural biotope of CMOS technology; Therefore it is well fit for CMOS integration.

- The increased reference frequency lowers the necessary division factor, such that the circuit noise in the PLL band is less amplified (Eq. (2.17)); As a result the in-band noise, and thus the rms phase error, can be low, even for fully integrated solutions.

- The broader bandwidth makes the synthesizer ready for future high data rate communication systems by allowing faster frequency switching.

- The $\Delta\Sigma$ fractional-N PLL requires only one integrated PLL and a digital $\Delta\Sigma$ modulator, reducing the power consumption and the occupied area.

Therefore, this book focuses on the monolithic integration of a $\Delta\Sigma$ fractional-N PLL frequency synthesizer which complies with the high-quality DCS-1800 specifications.

The design of a monolithic frequency synthesizer fits into a broader picture: the development of a fully integrated transceiver front-end for high-quality digital communication systems, shown

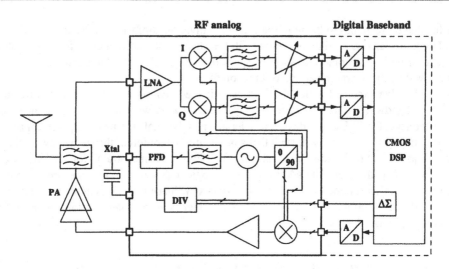

Figure 2.19: *A complete DCS-1800 cellular system can be built by flanking the transceiver IC by only a duplexer, a power amplifier, a Xtal oscillator and a digital baseband chip.*

in Fig. 2.19; By combining the $\Delta\Sigma$ fractional-N synthesizer with a low-IF receiver and a direct upconversion transmitter, the total front-end can be integrated on a single die. The transceiver front-end is targeted for half-duplex communication systems, i.e. transmission and reception never occurs at the same time. Therefore, the frequency synthesizer can be shared by the receive and the transmit path, leading to area and power savings. For the receiver a low-IF architecture has been chosen since it offers a high integratability by omitting interstage high-Q filtering. Additionally, it does not suffer from the DC offset and $1/f$ noise problems of the zero-IF receiver. The implementation details of the receiver can be found in [Jans02]. The transmit path is implemented by a direct conversion transmitter, which provides the most integrated solution, since it does not require high-Q interstage filtering. The implementation details of the transmitter are elaborately discussed in [Borr02]. A complete DCS-1800 cellular system can be built by flanking the transceiver IC by a minimum number of external components: a duplexer, a power amplifier, a Xtal oscillator and a digital CMOS baseband chip (see Fig. 2.19).

Coupling of digital noise through the substrate still stands in the way of an actual single-chip CMOS solution with the digital signal processing and the analog front-end on the same substrate. The presented front-end architecture allows for a two-chip solution. In that case, the ADCs and DACs should be integrated on the front-end chip, where they have a more quiet ground. The accuracy requirements for the ADCs are rather high (14 bit), but feasible in CMOS [Gee00] with a power consumption of around 10 mW. For reasons of digital noise coupling, the integration of the inherently digital $\Delta\Sigma$ modulator is assigned to the digital baseband chip. The design of the $\Delta\Sigma$ modulator is therefore not further discussed. The quadrature mixing to translate the 100 kHz IF signal to baseband and the actual image rejection is also performed in the digital baseband chip, together with the actual channel selection and digital AGC.

The main design goal of the transceiver front-end can be summarized as: "Complying with the stringent DCS-1800 specifications with an as high as possible degree of integration in standard CMOS technology, powered by a single 2V power supply while minimizing the overall power consumption".

2.6.2 The DCS-1800 Communication System

DCS-1800, a.k.a. GSM-1800, is a digital cellular communication system, using digital information coding and digital control. Handsets, further referred to as mobile stations (MS), roam throughout a honeycomb cell pattern, constantly keeping touch with the base transceiver station (BTS) in the center of the current cell. In what follows, only the part of the total DCS-1800 protocol that determines the specifications of the RF analog front-end of the MS is touched upon. Basically, DCS-1800 is the high frequency sibling of the well known Global System for Mobile communications at 900 MHz (GSM-900). The digital part of the standard, e.g. the communication protocol, the modulation method, the DSP algorithms, ..., is identical. The difference lies in the requirements for the RF front-end, i.e. the physical layer of the DCS-1800 protocol.

The frequency plan for the DCS-1800 mobile station consists of two bands; the frequency band from 1710 MHz to 1785 MHz is used for transmission, while the band from 1805 MHz to 1880 MHz is assigned to reception. DCS-1800 is a frequency-division-multiple-access (FDMA)/ time-division-multiple-access (TDMA)-based system where the physical channels are orthogonal in frequency and time. Each band of 75 MHz is subdivided in 373 channels, with a carrier spacing of 200 kHz, meaning that each communication channel is 200 kHz wide (FDMA). Potentially, up to 8 /users can share one channel, meaning that each channel is subdivided in 8 time slots or *burst periods*, which form one TDMA frame. Each mobile station is dynamically assigned a carrier frequency and a time slot through which the communication takes place. The receiver and transmitter of the MS are active approximately one-eighth of the time but never at the same time (half duplex), which means they can share the frequency synthesizer. Fig. 2.20 shows the timing of traditional DCS-1800. The monitoring frames provide time for the receiver to monitor the signal strength of the neighboring cells to provide information for hand-over and cell re-selection. For the hand-over to neighboring cells, special frames are allocated to synchronize the surrounding cells from time to time (not shown in Fig. 2.20).

13 frames occupy exactly 60 ms, such that each frame is approximately 4.615 ms long. Since one frame consists of 8 time slots, one time slot has a duration of 577 μs. Each of these time slots contains 156.25 bits – a 148 bit frame, with 2×57 information bits and 8.25 additional bits –, leading to an overall bit rate of 270.833 kbps. The specification for GSM supports a maximum data rate of 14.4 kbps. To increase the data rate, multiple time slots per frame can be allocated for one connection; This is done in HSCSD (High-Speed-Circuit switched data) and GPRS (General-Packet-Radio Service), providing data rates up to 57.6 kbps and 171.2 kbps respectively.

The occupied bandwidth for the 270 kbps data rate is restricted to the 200 kHz channel by pulse shaped modulation, i.e. GMSK modulation with a filter-bandwidth-bit-rate product of 0.3. GMSK (Gaussian Minimum Shift Keying) is a smoothed form of MSK, which is in fact binary FM. The data is coded in the instantaneous frequency rather than in the phase (as in e.g. QPSK).

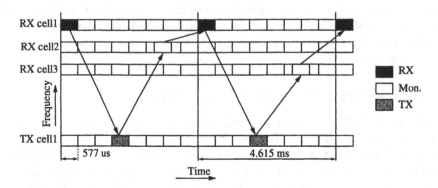

Figure 2.20: *The timing of traditional DCS-1800 without frequency hopping.*

Both GMSK and MSK are constant envelope modulation techniques, enabling high power-added efficiency power amps. Although the GMSK spectrum is more compact than MSK, the leakage into the adjacent channels is rather large. Therefore, adjacent channel frequencies are often not used at the same time, increasing the channel spacing to 400 kHz. The typical shape of the GMSK spectrum can be viewed in Fig. 2.23. The newest high data rate GSM flavor, the Enhanced Data rate for GSM Evolution (EDGE) standard, uses 8PSK (Phase Shift Keying) instead of GMSK to boost the performance of HSCSD and GPRS to data rates up to 400 kbps (ECSD and EGPRS) with a penalty of lower area coverage and a non-constant envelope modulation [EDG99].

2.6.3   From DCS-1800 to Synthesizer Specifications

The DCS-1800 standard [ETSI00] specifies the minimum reception and transmission quality that must be guaranteed under a set of given conditions, e.g. adjacent channel interference, blocking signals, small input signals, intermodulation requirements, ... In this section, the DCS-1800 specifications are systematically mapped onto measurable specifications of the frequency synthesizer, which provides the local oscillator signal for a low-IF receiver and a direct conversion transmitter. The reception quality for the smallest signals is reflected in a required *BER*, which determines the phase noise performance of the frequency synthesizer at higher offset frequencies. The same *BER* requirement also determines the maximum level of the spurious tones in the output spectrum of the frequency synthesizer. The transmission quality requirement sets the phase accuracy of the local oscillator signal, i.e. the rms phase error specification and the spurious suppression. The required maximum data throughput of the transceiver sets the settling constraint.

2.6.3.1   From Bit-error Rate to Signal-to-Noise Ratio

The sensitivity and the quality of the reception in digital communications systems is defined in terms of the bit-error rate (*BER*), i.e. the percentage of bits that are erroneous relative to the total

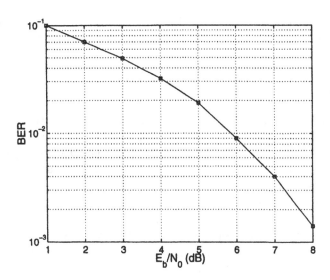

Figure 2.21: *The bit-error rate of a GMSK signal as a function of E_b/N_0, the SNR for a noise bandwidth equal to the bit rate.*

	Reference Sensitivity	Maximum BER
Class I/II	-100 dBm	2%
Class 3	-102 dBm	2%

Table 2.1: *The reference sensitivity performance for the DCS-1800 mobile station.*

number of bits received. In order to derive specifications for the RF analog front-end building blocks of a total digital transceiver, the required *BER* is mapped onto an equivalent minimum signal-to-noise ratio (*SNR*) [TSB98]. For a given (de)modulation method, the *BER* can directly be translated to a corresponding *SNR*. In Fig. 2.21, the theoretical *BER* of a GMSK modulated signal is plotted versus E_b/N_0, the *SNR* of the signal in a bandwidth equal to the bit rate.

In Table 2.1, the reference sensitivity performance of a DCS-1800 mobile station [ETSI00, p.35, 6.2 & p.27], i.e. the reference sensitivity and the corresponding maximum allowed *BER*, is listed. The reference sensitivity is defined as the minimum signal level that is still detectable with a given *SNR*, depending on the application. The reference sensitivity performance is given for all three DCS-1800 classes, from which the class 3 is clearly the most sensitive; In a class 3 mobile station, a *BER* of 2% needs to be achieved for input signal powers as low as -102 dBm. The class 3 mobile station requirements are therefore taken as a worst-case reference for the derivation of building block specifications. Using Fig. 2.21, the class 3 *BER* requirement translates to a *SNR* of approximately 4.9 dB in a bandwidth equal to the bit rate, which is 270.83 kbps during a DCS-1800 burst. The channel spacing of DCS-1800 is 200 kHz, such that the required *SNR* in a communication channel becomes 6.2 dB. To derive the specifications of the DCS-1800 frequency synthesizer, an *SNR* of 7 dB is targeted for -102 dBm input signals, which results in a *BER* of 1.2% for class 3 mobile stations. Equivalently, an *SNR* of 9 dB is targeted for -100 dBm input signals, corresponding to a *BER* of 0.2% for class I/II mobile stations.

2.6.3.2 Phase Noise

According to [ETSI00, p. 29, 5.1 & p. 35, 6.2], a wanted signal *3 dB above* the reference sensitivity level of -102 dBm must still be detectable with a *BER* of 2% in the presence of large, *sinusoidal* blocking signals. The maximum power of these sinusoidal blocking signals is shown in Fig. 2.22 as a function of the frequency offset from the wanted channel. As the out-of-band blockers are strongly attenuated by the antenna filter-duplexer, the largest remaining blocking signals are the blockers in the receive band.

If the local oscillator signal were ideal, the only requirement would be that the signal path does not saturate when a blocker propagates through the receiver and that the A/D converters have a large enough dynamic range to sample the wanted signal together with the blocker. However, in practice the phase noise skirt of the carrier (see Fig. 2.22) downconverts the blockers onto the wanted signal, degrading the signal's *BER*. In the same way, the adjacent channels are downconverted onto the wanted signal. Fig. 2.23 shows the power levels of the adjacent channels with respect to the wanted signal [ETSI00, p. 37, 6.3]. The GMSK power spectra are drawn for an integration bandwidth of 30 kHz as specified.

The most critical unwanted signals that can be downconverted onto the desired signal are the adjacent channel at 400 kHz (maximally 41 dB larger than the wanted signal), the -43 dBm blocking signal at 600 kHz, the -33 dBm blocking signal at 1.6 MHz and the -26 dBm blocking signal at 3 MHz. A *SNR* of 7 dB must be obtained after down-conversion (see the previous section), for a signal *3 dB above* the reference sensitivity, which is -102 dBm+3 dB $= -99$ dBm for class 3 DCS-1800 (see Table 2.1). Keeping in mind that one channel is 200 kHz and that the phase noise is white-like on a 200 kHz scale, the phase noise specification becomes:

$$\mathcal{L}\{400\,\text{kHz}\} \leq -41\,\text{dB} - 7\,\text{dB} - 10\log(200\,\text{kHz}) = -101\,\text{dBc/Hz}$$

$$\mathcal{L}\{600\,\text{kHz}\} \leq -99\,\text{dBm} - (-43\,\text{dBm}) - 7\,\text{dB} - 10\log(200\,\text{kHz}) = -116\,\text{dBc/Hz}$$

$$\mathcal{L}\{1.6\,\text{MHz}\} \leq -99\,\text{dBm} - (-33\,\text{dBm}) - 7\,\text{dB} - 10\log(200\,\text{kHz}) = -126\,\text{dBc/Hz}$$

$$\mathcal{L}\{3\,\text{MHz}\} \leq -99\,\text{dBm} - (-26\,\text{dBm}) - 7\,\text{dB} - 10\log(200\,\text{kHz}) = -133\,\text{dBc/Hz} \quad (2.47)$$

Assuming that the phase noise of a frequency synthesizer has a -20 dB per decade roll-off for the critical offset frequencies, the phase noise specification at 3 MHz is the most stringent.

2.6.3.3 Spurious Suppression

The spurious suppression specification for the DCS-1800 standard is set by transmission as well as reception requirements. For transmission, the synthesizer output spectrum must comply to the power mask in [ETSI00, annex A], such that after upconversion, the transmitted signal is within the power mask. In Fig. 2.24, the transmission determined spurious suppression is plotted

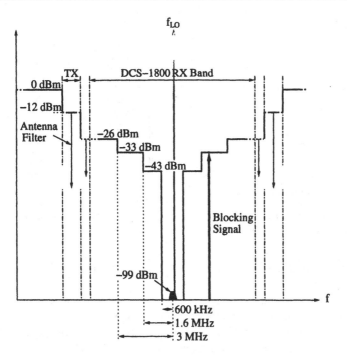

Figure 2.22: *Maximum blocking signal power as a function of the frequency offset from the wanted channel.*

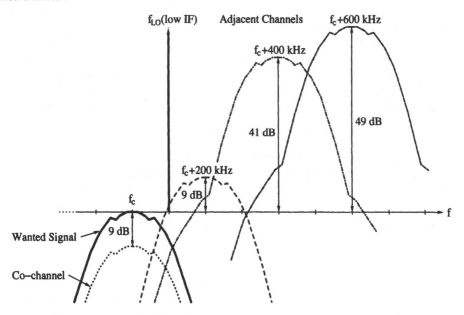

Figure 2.23: *Co-channel and adjacent interference in DCS-1800.*

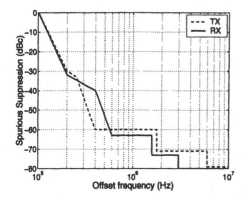

Offset (kHz)	TX (dBc)	RX (dBc)
200	-30	-32
400	-60	-48
600-1600	-60	-63
1600-1800	-60	-73
1800-3000	-71	-73
3000-6000	-71	-80
≥ 6000	-79	-80

Figure 2.24: *The spurious suppression specification for DCS-1800 at different offset frequencies for transmission (TX) and reception (RX).*

(dashed) and the numbers [ETSI00, p.16, 4.2] are given in the table for the worst-case TX output power.

For low-IF reception, the spurious suppression is determined by the same specifications that set the phase noise. A spurious tone at a certain frequency offset from the carrier can act as a pirate local oscillator signal for blocking signals and adjacent channels. Again, under all conditions a -99 dBm wanted signal must be detectable with a 7 dB *SNR*. A spurious tone at 200 kHz downconverts the adjacent channel onto the wanted channel. Since the adjacent channel is 9 dB higher than the wanted signal, the spurious suppression is only -7 dB $- 9$ dB $= -16$ dBc. However, the same 200 kHz spur also downconverts the tail of the following adjacent at 400 kHz (see Fig. 2.23) onto the 200 kHz adjacent channel. To satisfy the spec, the downconverted tail must be as low as the tail of the 200 kHz adjacent channel, i.e. suppressed by 32 dB. A spurious tone at 400 kHz downconverts the second adjacent channel onto the wanted channel, such that the spurious suppression is -7 dB $- 41$ dB $= -48$ dBc. For spurious tone frequencies higher than 600 kHz, the specification is determined by the blocking signal power levels. From 600 kHz to 1.6 MHz, the possible blocker is -43 dBm (see Fig. 2.22). The resulting suppression is then:

$$-99 \text{ dBm} - 7 \text{ dB} - (-43 \text{ dBm}) = -63 \text{ dBc} \qquad (2.48)$$

The suppression at offset frequencies corresponding to the other blocking levels is calculated in the same way and summarized in Fig. 2.24. For offset frequencies higher than 3 MHz, i.e. for the reference spur at 26 MHz, the spurious suppression needs to be less than -80 dBc. This is however a worst-case value; When the reference spur is outside the RX band, the signal levels that can be downconverted by this spur are seriously reduced by the antenna-duplexer filter.

2.6.3.4   rms Phase Error

The phase accuracy of the synthesizer is determined by the transmission characteristics of the DCS-1800 standard [ETSI00, p.20, 4.6]. The phase error is measured by computing the difference between the phase of the transmitted waveform and the phase of the expected one. The

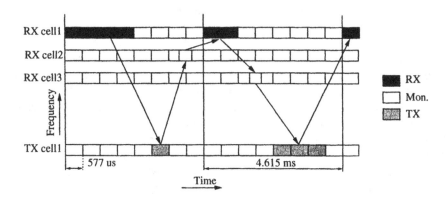

Figure 2.25: *The timing for GPRS multi-slot class 12 for different multi-slot allocation of the 5 possible time slots per frame.*

GMSK modulation requires a phase error of less than $5°$ rms with a maximum peak deviation of $20°$. The total rms phase error of a MS, $\Delta\Phi_{tot}$, is calculated by taking into account the four main contributors:

$$\Delta\Phi_{tot}^2 = \Delta\Phi_{PLL}^2 + \Delta\Phi_{BB}^2 + \Delta\Phi_{mix}^2 + \Delta\Phi_{PA}^2 \qquad (2.49)$$

with $\Delta\Phi_{PLL}$ the rms phase error of the frequency synthesizer, $\Delta\Phi_{BB}$ the rms phase error generated in the baseband modulation, $\Delta\Phi_{mix}$ the rms phase error of the RF mixers and $\Delta\Phi_{PA}$ the rms phase error of the power amplifier, due to saturation and AM-to-PM conversion. To enable some implementation margin for all the contributors, the rms phase error of the synthesizer should be restricted to $2°$ or less. Using Eq. (2.24) for a PLL noise bandwidth of e.g. 35 kHz, this corresponds to an in-band phase noise floor of -78 dBc/Hz. The synthesizer output *SNR* is then 26 dB (see Eq. (2.26)).

2.6.3.5 Dynamic Performance

The settling time is specified by the DCS-1800 standard [ETSI00] for a frequency step of 95 MHz with a frequency accuracy of 180 Hz. This frequency step must be achieved in 865 μs. This requirement comes from the need to synchronize to one of the surrounding cells with the monitoring frames. To enable monitoring the synthesizer must be able to settle within 1.5 time slots, i.e. 865 μs (see Fig. 2.20).

For the high-data-rate standards, HSCSD, GPRS and their EDGE counterparts, ECSD and EGPRS, the timing situation is totally different and much more stringent. The high data rates are achieved by using multiple time slots within one frame for data communication. 29 multi-slot classes are defined [ETS00, Annex B], the most stringent of which is multi-slot class 12 under the constraint of a type 1 mobile station, i.e. with a common frequency synthesizer for the RX and TX path. The multi-slot class 12 standards allocates 5 time slots for data transfer within one frame with maximally 4 TX or 4 RX time slots (see Fig. 2.25).

Two factors limit the switching time of the frequency synthesizer; First, when switching from a receive to a transmit time slot, the *timing advance* must be taken into account. The timing advance is the time the MS has to transmit earlier to the BTS to be in the timing schedule of the BTS, depending on the distance between them. The maximum timing advance the BTS can request is 233 μs, corresponding to 35 km. A second factor is the time needed for the adjacent cell signal level measurement. Depending on the quality of the DSP, this time is between 150 and 300 μs. To compute the switching time the monitor burst is assumed to be centered in the middle of the gap between successive RX and TX bursts.

The settling time of the DCS-1800 frequency synthesizer can be calculated to be 310 μs for GPRS and 288 μs for HSCSD, due to additional synchronization issues. Although HSCSD provides a lower data rate, the settling requirements for the synthesizer are more stringent than for GPRS. Using Eq. (2.33), the bandwidth of a first-order loop must at least be 7.3 kHz.

2.6.3.6 Specification Summary

In Table 2.2, a summary is given of the specifications for the frequency synthesizer in a fully integrated transceiver for traditional DCS-1800 and its high-data-rate siblings, HSCSD, GPRS, ECSD and EGPRS. The spectral purity specifications are derived for an overall *SNR* of 7 dB and for input signals 3 dB above the reference sensitivity level. Power consumption and area is specified as low as possible (ALAP). The supply voltage constraint is chosen to enable future integration of the digital and the analog part of the front-end on one die with the same power supply voltage. Some margin is taken into account to ensure all specifications.

2.7   Conclusion

To realize a complete transceiver for wireless communication systems, one indispensable building block is required by both the receive and transmit path, i.e. the building block that generates the local oscillator (LO) signal: *the frequency synthesizer*. The aim of this book is to implement a frequency synthesizer that satisfies the same constraint imposed on the receiver and transmitter: cost reduction. The phase-locked loop frequency synthesizer has been identified as the best solution, since it has the potential of combining high frequency and low power operation and is very well suited for integration in today's most cost-effective IC technology: CMOS.

To start the analysis of PLL frequency synthesizers, the synthesizer data sheet is drawn up, starting with the most critical static specification: the spectral purity. The definition of phase noise and how it behaves throughout the PLL is elaborated, out-of-band as well as in-band, i.e. the rms phase error. Additionally, spurious suppression determines the spectral purity of the PLL. The dynamic specifications are given by how fast and with what error the PLL tracks its input and in which range it can acquire phase-lock. More general specifications include tuning range, frequency resolution, power consumption and integratability.

The different specifications are mapped onto the different building blocks of the PLL frequency synthesizer. The PLL synthesizer is mainly a filtering feedback system, consisting of a phase detector which compares the input with the output, yielding an error signal which is fil-

Phase Noise		Section 2.6.3.2
at 600 kHz offset	-116 dBc /Hz	
at 3 MHz offset	-133 dBc /Hz	
Spurious Suppression		Section 2.6.3.3
Reference Spurs	< -80 dBc	
Fractional Spurs	see Fig. 2.24	
rms Phase Error $\Delta\Phi_{rms}$	2°	Section 2.6.3.4
Settling Time (95 MHz /180 Hz)		Section 2.6.3.5
DCS-1800	865 μs	
(E)GPRS	310 μs	
(E)(HS)CDS	288 μs	
Frequency Resolution	200 kHz	Section 2.6.2
Center Frequency	1.8 GHz	Section 2.6.2
Tuning range	170 MHz	Section 2.6.2
Power Consumption	*ALAP*	Section 2.6.1
Area	*ALAP*	Section 2.6.1
Supply Voltage	1.8V-2V	Section 2.6.1

Table 2.2: *Summary of the frequency synthesizer specifications in a DCS-1800 communication system.*

tered by the loop filter and applied to the oscillator for frequency control. Optionally, a frequency divider is implemented to multiply the reference frequency to the output. All different building blocks and their different existing implementations are discussed in detail, with the focus on the respective importance on the different specifications listed in the synthesizer data sheet.

To demonstrate the feasibility of high-quality, monolithic frequency synthesizers in CMOS technology, the stringent specifications of the DCS-1800 standard are set as a design target; The DCS-1800 communication system and the intended front-end integration are elaborated, with a focus on the integration of the synthesizer. It has been concluded that a simple integer-N PLL is unable to satisfy the demands of DCS-1800. More advanced synthesizer topologies need to be explored, preferably $\Delta\Sigma$ fractional-N PLLs since they trade analog accuracy for digital complexity, the biotope of CMOS. Using the DCS-1800 as a demonstration vehicle, general DCS-1800 requirements are mapped onto the frequency synthesizer data sheet, which can then further be projected onto PLL building block specifications, providing the leitmotiv of the book.

Chapter 3

High-Speed CMOS Prescalers

3.1 Introduction

As mentioned in the introduction on frequency synthesis in Chapter 2, the two building blocks operating at high frequencies in a PLL frequency synthesizer are the VCO and the frequency divider. This chapter covers the design and implementation of high-speed frequency dividers. The design of low-phase-noise, high-speed VCOs is covered in the following chapter. Frequency dividers are usually implemented as digital counters. The speed of digital counters is however limited in CMOS technology. Therefore, high-speed prescalers are used to pre-divide the frequency. As discussed in Section 2.4.4.1, a fully programmable, high-speed frequency divider can be realized with a dual-modulus prescaler and two digital counters.

The main trade-off in prescaler design is between speed and power; The prescaler is actually a high-speed digital circuit, implying that its operation follows the dynamic power law of digital circuits: $P \propto C_L \cdot f \cdot V_{DD}^2$. As indicated by the dynamic power law, the prescaler can benefit from the continuous scaling of CMOS technology (lower C_L) and supply voltage V_{DD}. The constant of proportionality is an indication for the quality of the design. Depending on the application, the input sensitivity, i.e. the smallest input signals the prescaler can handle, the residual phase noise and the nature of the input signal, i.e. single-ended or differential, are of importance to characterize the performance of a prescaler.

In this chapter, the optimization of prescaler design for different applications is elaborated with four implementation examples. The main target in all designs is the speed-power optimization. Therefore, programmable division in the prescaler is implemented using the phase-switching topology, developed in [Cran98]. The phase switching topology allows to exploit the full speed of a single divide-by-2 flipflop. The phase-switching technique and its design issues, especially the possibility of spurious generation, are discussed in more detail.

The first two prescaler implementations are both dual-modulus prescalers for use in discrete, state-of-the-art frequency synthesizers. Discrete synthesizer implementations usually integrate the digital synthesizer part on one chip, while the high-speed blocks are on separate ICs. The output of most commercially available VCOs is single-ended and small to minimize LO feedthrough. Therefore, the dual-modulus prescaler has to offer high-speed operation with a very high input sensitivity . Moreover, the phase noise of commercial VCO can be extremely low through the use of exotic techniques to realize high Q inductors. As a consequence, the residual phase noise of the prescaler can become an issue. Therefore, phase noise generation and minimization in prescaler circuits is investigated and applied in prescaler design. Two dual-modulus prescaler prototypes are presented for discrete, state-of-the art frequency synthesizers that achieve GHz operation with small, single-ended input signals in rather ancient standard CMOS technology, a 0.7 μm CMOS and 0.8 μm "radiation hardened" BiCMOS process.

The subsequent implementation is a 16-modulus prescaler for use in a fully integrated CMOS frequency synthesizer for DCS-1800. The multi-modulus operation is necessary to enable division in the full frequency range of the VCO. The multi-modulus operation is realized by extending the phase-switching process to multiple swallow operations by additional control signal processing. Since the prescaler is part of a fully integrated PLL, its input signal, provided by the on-chip VCO is differential and almost rail-to-rail. Moreover, the prescaler residual phase noise is drowned in phase noise coming from the integrated loop filter and is therefore not critical. The design focus is on the speed-power trade-off and on robustness, while satisfying the DCS-1800 requirements (Table 2.2). The imposed tuning range translates to division moduli between 65 and 73 for a 26 MHz reference frequency.

Last but not least, a prescaler in the true sense of the word, i.e. with a fixed division modulus, is developed to explore the speed limits of prescaler design in standard CMOS technology. A 12 GHz divide-by-128 prescaler is integrated in a first-generation 0.25 μm technology. The implementation combines high speed, high input sensitivity and multi-stage division in CMOS, making it one of the fastest, functional CMOS prescaler implementations published in open literature.

3.2   The Phase-Switching Dual-Modulus Prescaler

3.2.1   Conventional Architecture

The conventional implementation of a dual-modulus prescaler is depicted in Fig. 3.1 for a 64/65 division. The conventional dual-modulus prescaler usually consists of a divide-by-4/5 synchronous divider and a divide-by-16 asynchronous divider, which is no more than a chain of four D-flipflops. When the mode input M of the prescaler is zero, the output of the second NAND gate in the synchronous divider is always logic one. Consequently, the first NAND gate acts as a simple inverter and the first two D-flipflops form a synchronous divide-by-4 frequency divider. Together with the D-flipflop chain, the total division modulus of the prescaler is 64. When the mode input of the prescaler is logic one, the input to the second NAND gate becomes one for a brief moment once every output period of the prescaler. During this moment, the output of the second NAND gate is the inverted output of the second D-flipflop, such that the feedback loop is closed over the three D-flipflops of the synchronous divider. The third flipflop adds an extra delay of exactly one period of f_{in}. This delay is added to the output of the prescaler once every output period, resulting in a division by 65.

The main speed bottleneck of the conventional dual-modulus prescaler architecture is situ-

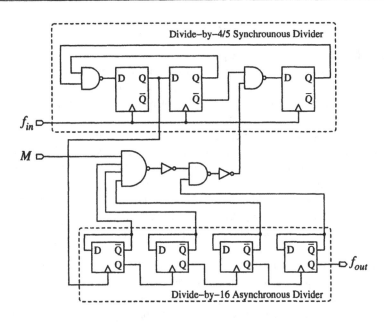

Figure 3.1: *A conventional divide-by-64/65 dual-modulus prescaler.*

ated in the synchronous divider. All three fully functional D-flipflops of the synchronous divider operate at the highest frequency, seriously increasing the clock load and the power consumption of this architecture. What's even worse, both NAND gates are in the critical path of the divider; The additional delay causes the maximum operating frequency of the conventional prescaler architecture to be much lower than that of a single high-speed D-flipflop. Moreover, the synchronous divider requires fully functional D-flipflops while a single D-flipflop can be optimized for divide-by-two operation [Huan96]. Publications in the field report speed degradations ranging from 2.8 GHz for the D-flipflop to 2 GHz for the full prescaler [Cong88] and 2.25 GHz down to 1.22 GHz for the full prescaler [Chan96].

One way to improve the speed bottleneck is embedding the NAND logic in the D-flipflop as in [Rog94]. The reported speed degradation is only from 1.57 GHz to 1.4 GHz. To further improve the power consumption of the synchronous divider, [Kado93] reduces the number of high-speed D-flipflops by realizing a divide-by-2/3 divider instead of 4/5. The power consumption is reduced by up to 20%, but the operation margin of the prescaler is halved. While the previous ideas are improvements on the conventional architecture, others searched for ways to implement the dual-modulus function with the same speed as a single divide-by-2 flipflop. The approach proposed in [Lars96] is to pre-process the input signal and than use a cascade of asynchronously connected divide-by-2 stages for high-speed operation; When the mode input is logic one, every output period, one input period is blocked from the divider chain, resulting in $N + 1$ division. The dual-modulus prescaler has the same speed as a fixed asynchronous divider chain. Another architecture that enables to exploit the full speed performance of a single flipflop is the phase switching architecture [Cran98].

3.2.2 The Phase-Switching Architecture

The phase-switching architecture [Cran98] is illustrated in Fig. 3.2 (a) for a 64/65 dual-modulus prescaler. The basic architecture consists of an asynchronous chain of divide-by-2 flipflops. The dual-modulus function is realized by the phase-select block situated after the second divide-by-2 flipflop. This architecture exploits the full speed of a single D-flipflop, since no additional logic is needed in the high-speed critical path. Moreover, the high-speed D-flipflop does not have to be fully functional and can therefore be optimized for divide-by-2 operation; The input is always 01010101... and the output is always 00110011... (relative to the input frequency). As shown in Fig. 3.2 (a), the phase-select block operates on 90° shifted output signals of the second flipflop. By implementing the second flipflop as a master/slave flipflop, these quadrature signals are inherently present; The differential master output is denoted by $F4.I$ and $\overline{F4.I}$, while the 90° shifted slave output is denoted by $F4.Q$ and $\overline{F4.Q}$. When the mode input of the prescaler is zero, the mode-NAND gate outputs a one and the phase control block is inactive. In this case, the single-ended input frequency f_{in} is divided by two, to reveal the differential input $F2$ and $\overline{F2}$ for the half-speed M/S flipflop. The phase select block passes the currently selected output phase of the M/S flipflop to the output $F4$, which is further divided by a low-speed asynchronous divide-by-16 divider. When the mode input M is logic one, the mode-NAND operates as an inverter and the phase control blocks starts generating control pulses for the phase-select block. To illustrate the dual-modulus operation, the waveforms in the prescaler are plotted in Fig. 3.2 (b). Every rising edge of the prescaler output signal f_{out}, the phase-select block selects the input that lags its current input by 90°; As a result, the phase-select output $F4$ is delayed by 90°, as shown in Fig. 3.2 (b) for a switch between $F4.I$ and $F4.Q$. A delay of 90° on the the input signal f_{in} divided by 4, corresponds to a delay of one full period of the input signal. The full period delay is added every period of the output signal, hence a division by 65 is accomplished.

A possible drawback of the phase-switching architecture is situated in the transition from one phase to another (see Fig. 3.2 (b)). The transition must occur smoothly under all circumstances, since a spike in the transition would trigger the divide-by-16 block leading to a erroneous overall division. To understand the occurrence of spikes, the phase-select block and the corresponding control by the phase control block is discussed in more detail (see Fig. 3.3). Assuming that the outputs of the half-speed divide-by-2 flipflop are nearly rail-to-rail signals with fast transitions, the phase-select block can be implemented by purely digital circuitry. In this case, the selection between two phases is performed by three NAND-gates, which form a selection block in Fig. 3.3. The control signals $C1$ and $C2$ control the differential switching, while $C0$ controls the actual 90° phase transition between $F.I$ and $F.Q$. The control signals are generated in the phase control block, which is a small finite-state machine that ensure the right relative timing of the signals. The timing of the differential switching is such that it occurs only when the corresponding phases are not connected to the output, i.e. if $F.I$ is connected to $F4$, $C2$ changes state. Therefore the differential switching is not critical for spike generation. The possible parasitic spike generation occurs when $C0$ changes state, especially at low operating frequencies, since at high frequencies the NAND gates are too slow to respond to fast control signal changes. The spike generation effect is dependent on three parameters: the time of the arrival and the slope of a change of $C0$ and the delay between $C0$ and $\overline{C0}$. If for example $C0$ undergoes a fast low-to-high transition,

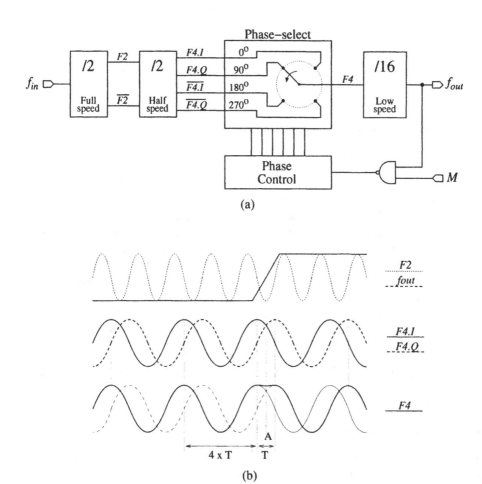

Figure 3.2: (a) A divide-by-64/65 dual-modulus prescaler based on the phase switching principle of [Cran98]. (b) The phase-switching principle illustrated with waveforms.

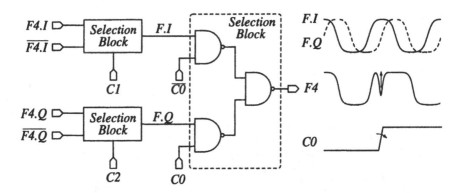

Figure 3.3: *The phase-select block circuit with the corresponding control signals.*

before time A in Fig. 3.2 (b), $F4$ becomes erroneously low, since $F.Q$ is still low at the moment, creating a spike (see Fig. 3.3).

Re-timing of the control signals to control the time of arrival is hard, since the delay through the circuit is not well known and dependent on process and temperature variations. In all presented implementations, $C0$ and $\overline{C0}$ are generated by a differential flipflop to minimize their relative delay. The timing of the control clock is such that under typical conditions the timing of $C0$ is ideal. To patch the timing problem, the slope of the transitions of $C0$ has been lowered by implementing small control buffers with a fixed additional output capacitance. Simulations reveal that the range over which the slope may vary without creating spikes is very large. The control buffers are designed such that they guarantee a smooth phase transition under temperature and process variations and different arrival times of the control signal.

3.3 A Single-Ended 1.5 GHz 8/9 Dual-Modulus Prescaler in 0.7μm CMOS

3.3.1 The High-Speed Divide-by-2 D-flipflop

As mentioned earlier, the high-frequency behavior in the phase-switching architecture entirely depends on the first divide by-2 flipflop. Several topologies are available for implementation in CMOS. To go to very high frequencies, one can use the CML-like (Current Mode Logic) D-flipflop, by combining two latches of Fig. 3.4 (b), which exhibits a very good speed-power trade-off. However, the CML-like flipflop operates with a differential input signal. This means that for single-ended applications a high-frequency single-ended to differential converter has to be used, which seriously increases the power and degrades the residual phase noise. Moreover, the division operation of CML-like flipflops is highly sensitive to the DC input level of the clock ϕ; A DC shift of only a few tens of mV can be sufficient to cause the circuit not to divide at all. The high operation speed of the CML-type flipflop is achieved by operating with small internal signals, enabling fast state changes. This is however detrimental for the phase noise performance

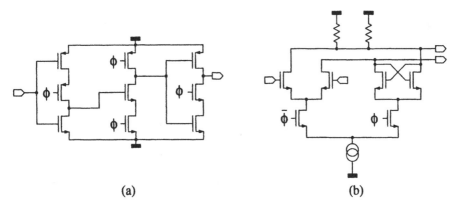

(a)   (b)

Figure 3.4: *(a) A conventional TSPC nine-transistor D-flipflop and (b) an CMOS version of the bipolar CML-latch*

of the circuit (see Section 3.3.3).

On the other hand, TSPC (True-Single-Phase-Clock) dynamic D-flipflops [Yuan89] (see Fig. 3.4 (a)) can combine both single-ended and high-frequency operation. The disadvantage of this type of circuits is that high frequencies can only be obtained with almost square-wave rail-to-rail input signals. However, most of the commercially available VCOs output a small signal. To overcome this problem most of the previously published prescalers apply input buffering [Chan96, Rog94], which consumes more power than the total prescaler itself.

To combine high-frequency, single-ended operation with very small input signals, an improved type D-flipflop is developed in this book. The circuit is shown in Fig. 3.5 and is derived from the edge-triggered, nine-transistor D-flipflop of [Yuan89]. Sizing of digital sequential circuits is not straightforward; Due to the large signal, time-varying nature of the transient signals, the behavior is almost impossible to express in closed formulas. Moreover, most of the transistors are both driver and load at the same time or in different states, which causes contradictory optimizing constraints. Therefore, the sequential circuit analysis technique presented in [Huan96] is adopted. Although this technique is intended for large input signals, sizing optimization rules can be extrapolated to small input signals. The technique helps to determine the relative sizing of the different transistors, but the absolute transistor sizing for a certain speed constrained is obtained from HSpice simulations. The D-flipflop is optimized for divide-by-2 operation.

First, the clock (ϕ) transistors are placed as close to the AC ground as possible. This offers a speed enhancement of around 20% [Huan96]. Another advantage of this approach is that the capacitance of the clock transistors is not in the critical path of the D-flipflop. As a result, the size of the clock transistor can be increased without speed limitation, resulting in an enhanced input sensitivity. In addition, an extra transistor Mp1 is added in the pre-charge stages of the D-flipflop. Due to transistor Mp1, ① is conditionally pre-charged and not every time the input changes from 1 → 0. This causes an activity decrease of ① and thus extra power savings. The extra transistor also increases the speed; When the input is low and the input of the pre-charge stage is logic 1, charge sharing through Mn1 causes a decrease of the voltage level at ①. When the

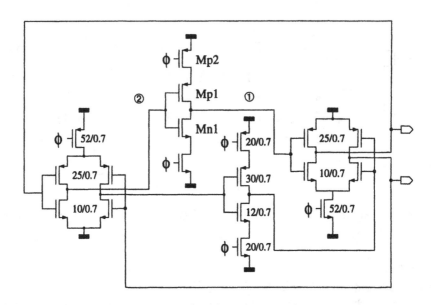

Figure 3.5: *The high-speed divide-by-2 single-ended to differential D-flipflop of the 0.7μm CMOS DMP.*

input changes from $0 \rightarrow 1$, ① discharges faster. The extra load presented by transistor Mp1 on ② is negligible, although its a relatively large transistor (see Fig. 3.5); Its capacitance is in series with the junction capacitance of clock transistor Mp2, when the latter is in the off-state ($\phi=1$). Consequently, the capacitance seen from ② is a lot smaller than the actual capacitance presented by Mp1. Another advantage of the D-flipflop is that, although it has a single-ended input, the circuit internally works fully differential. The outputs are perfectly differential, which cannot be realized with an inverter at the end of a single-ended output. Due to the differential operation, residual phase noise caused by power supply line and substrate perturbations is suppressed in the high-speed D-flipflop as well as in the next divide-by-2 flipflop.

The layout is optimized towards high speed by using multiple contacts and vias for resistance minimization and using multiple fingers for wide transistor to decrease the gate RC delay. Transistors stacked in a NAND-like structure share the drain/source area to minimize the gate separation and thus the junction capacitance. Interconnects are made as short as possible for minimal propagation delay.

3.3.2   The Half-speed Divide-by-2 D-flipflop

The circuit of the half-speed D-flipflop is shown in Fig. 3.6. Due to the differential input two n-latches can realize a master-slave flipflop, which produces the quadrature signal necessary for the phase multiplexing in the phase select block. The latches are DSTC (Dynamic-Single-Transistor-

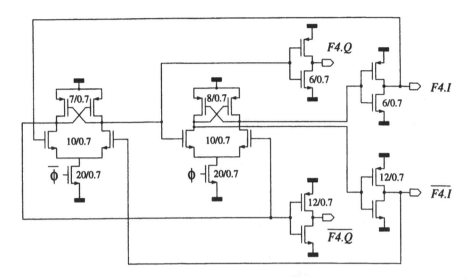

Figure 3.6: *The half-speed divide-by-2 differential to quadrature D-flipflop of the 0.7μm CMOS DMP.*

Clocked) n-latches [Yuan97]; As a result, only two clock transistors are necessary, minimizing the output load of the high-speed D-flipflop for optimal speed-power trade-off. The circuit is fully differential to maximize common mode noise suppression. The output inverters buffer the flipflop from the load of the phase select block, for optimal speed and power.

3.3.3 Phase Noise Considerations

The phase noise of a prescaler is most easily understood in the time domain; Since the prescaler is a chain of flipflops, noise present at the input of the flipflops or in the flipflop circuits modifies the triggering moment of the flipflop and causes time jitter. In Fig. 3.7 (a), this mechanism is illustrated for a voltage noise v_n that causes a time error $\Delta\tau$. The size of the time error for a constant voltage noise is inversely proportional to the slope of the waveforms in the prescaler circuit. Note that in an asynchronous divider chain, the jitter and hence the phase noise accumulates through the chain. In a synchronous divider, the output phase noise is only determined by the last flipflop. The jitter is related to phase changes in the circuit as follows:

$$\Delta\theta = \left.\frac{dt}{dv}\right|_{t=t_{trig}} \cdot v_n \cdot 2\pi f = \Delta\tau \cdot 2\pi f \tag{3.1}$$

with f the signal frequency and t_{trig} the triggering moment. Since the prescaler output is synchronized with the transitions of the prescaler input, the time jitter at the in- and the output is equal. Using Eq. (3.1), the output phase variation is related to the input phase variations in the same way as the input and output frequency, i.e. $\theta_{out} = \theta_{in}/N$. If the phase variations are

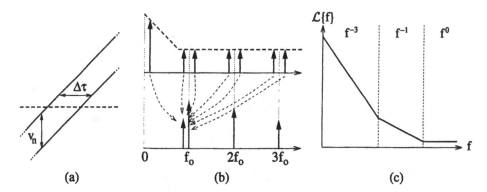

Figure 3.7: *Residual prescaler phase noise (a) in the time domain and (b) in the frequency domain and (c) a typical residual phase noise spectrum.*

represented by their phase power spectral densities to represent the prescaler phase noise in the frequency domain, the relationship becomes:

$$S_{\theta,out}(f_m) = \frac{S_{\theta,in}(f_m)}{N^2} \tag{3.2}$$

with f_m the offset frequency. In other words, the input phase noise of the prescaler is decreased from the input to the output by a factor N^2.

The prescaler also adds internally generated phase noise to the total output phase noise, which is referred to as the residual phase noise of the prescaler. Due to the feedback action of a PLL the total output phase noise power spectral density of the prescaler is multiplied by N^2. This indicates that residual prescaler phase noise can be important in frequency synthesizers. The residual phase noise in prescalers is generated by two different mechanisms [McCl92, Egan90]: additive and multiplicative noise. For both types of noise only the phase noise is of importance since digital circuits ideally suppress all amplitude variations. Additive phase noise corresponds to device noise (see the upper drawing in Fig. 3.7 (b)) and tends to be of the form

$$S_{\theta,add}(f_m) = K_0 + \frac{K_1}{f_m} \tag{3.3}$$

with K_0 and K_1 constants for white and flicker noise contributions, respectively.

Multiplicative noise originates from modulation of noise that is present around integer multiples of the output frequency due to the non-linear operation of flipflop circuits. The mechanism is depicted in Fig. 3.7 (b). In fact, noise around multiples of f_o is sampled by the flipflops and folds back around f_o creating the phase noise spectrum. The transfer function of device noise to the flipflop output is similar to that of a ring oscillator, since a divider is nothing more than an inverter chain locked to a certain frequency by the clock transistors. This means that for large offset frequencies, the transfer function is flat (i.e. the additive noise region), while for smaller offset frequencies the transfer function is proportional to $1/f_m^2$. As a result, the residual phase noise of a prescaler exhibits a noise floor at large offset frequencies and a $1/f_m$ related section for smaller offsets, i.e. determined by additive device noise. Due to modulation, white noise is upconverted to $1/f_m^2$ shaped noise as in oscillators. But since a divider is a locked oscillator, the white noise upconversion is greatly reduced and $1/f$ noise upconversion becomes dominant. This translates to a $1/f_m^3$ shaped phase noise region for small offset frequencies, which goes over to additive $1/f$ noise, skipping the $1/f_m^2$ region (see Fig. 3.7 (c)). This behavior is predicted by dedicated transient phase noise simulations performed on the presented prescaler in [DeSm98].

To capture the phase noise behavior in closed formulas is an endless task, due to the highly non-linear and sampled nature of the problem. Most reported attempts are purely empirical [Egan90, see refs: Kroupa and Robins]. Also simulating prescaler phase noise as in [DeSm98] is a tedious job, which is highly sensitive to the numerical accuracy of the simulations, since prescaler phase noise is inherently low due to the locking. Therefore, this book restricts itself to providing design guidelines for low-phase-noise prescalers.

- Dynamic TSPC D-flipflops are preferred, because their internal signals are rail-to-rail, whereas the internal signals in CML-type D-flipflops are much smaller (< 1V). As a result, dynamic D-flipflops have a larger signal-to-noise ratio, which means lower phase noise.

- As already mentioned, jitter is mainly introduced by noise on the triggering signals, causing a modification of the time at which the triggering threshold is crossed. When the slope of the triggering signals is large, the influence of noise is highly reduced. Dynamic CMOS circuits operate with very steep signals, because of their rail-to-rail nature. Therefore, phase noise can be expected to be lower than in CML-type flipflops, which operate with sine wave-like internal signals. This is confirmed by measurements presented in [McCl92].

- The D-flipflop of Fig. 3.5 operates internally fully-differential, resulting in improved noise suppression, compared to fully single-ended topologies.

3.3.4 Experimental Results

3.3.4.1 General Results

The presented prescaler is integrated in a 0.7μm CMOS technology. An IC microphotograph of the dual-modulus prescaler is shown in Fig. 3.8 with at the right side the output buffers to drive the signal off chip. The input signal is attenuated by the input bonding wire and the bonding pad capacitance. To boost the incoming signal the on-chip input is terminated with a 200Ω resistance, which causes an overshoot in the input transfer function in comparison with a 50Ω termination.

The measured maximum input frequency of the prescaler is 1.5 GHz using a 5V power supply. The output waveforms for divide-by-8 and divide-by-9 operation with a 1.5 GHz small input signal are shown in Fig. 3.9. The measured maximum input frequency and the corresponding power consumption are plotted versus the power supply voltage in Fig. 3.10 (a) for a fixed DC input. The prescaler operates at 1.5 GHz with a power consumption of 11mA drawn from a 5V power supply and at 820 MHz at 3V and 3.5mA. To get an idea of the performance of the circuit, the presented prescaler is compared to the 128/129 dual-modulus prescaler in [Cran98],

Figure 3.8: *IC Microphotograph of the dual-modulus prescaler in 0.7μm CMOS.*

which uses CML-like flipflops and is integrated in the same 0.7μm CMOS technology. The circuit consumes 8 mA at 3V for 1.75 GHz operation. The power consumption of CML type prescaler is higher due to the amplification that is needed to after the flipflops to drive the phase select block. The frequency-power ratio becomes 1750/24 = 72.9 MHz/mW, while this ratio is 820/10.5 = 78.1 MHz/mW for the presented implementation; This demonstrates the excellent speed performance of the presented TSPC prescaler, since its performance is comparable to that of an inherently faster CML-logic prescaler.

The power consumption of the prescaler is divided over the different building blocks as follows (in simulation) at 5V: High-speed DFF: 4.3mA, half-speed DFF: 2.7mA, phase-select: 1.7/3.2mA and the low-speed DFF: 1.5mA. The simulated total for divide-by-8 operation is 10.2mA which is close to the measured 11mA. The simulated maximum input frequency was more than 25% higher than the measured one, which is probably due to lack of high-frequency modeling of the transistors (HSpice level 2) and a higher than expected overlap capacitance in the actual processing.

In Fig. 3.10 (b), the required input signal level is presented. As can be seen the maximum input sensitivity over the whole frequency range is lower than 110mV$_{rms}$, taking into account the attenuation presented by the package parasitics. This is orders of magnitude smaller than other presented CMOS prescaler [Chan96, Rog94], without using input buffers. The high input sensitivity is achieved by implementing large clock transistors, which inevitably results in higher power consumption. However, the slightly increased power consumption is much less than the power consumed by high-frequency input buffers.

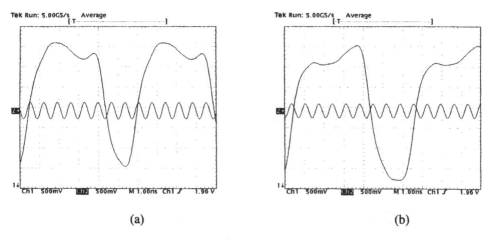

(a)   (b)

Figure 3.9: *In- and output waveforms for the DMP in 0.7 μm CMOS for (a) divide-by-8 and (b) divide-by-9.*

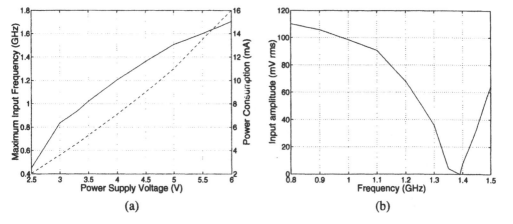

(a)   (b)

Figure 3.10: *Experimental results for the DMP in 0.7 μm CMOS: (a) The maximum input frequency and power consumption (dashed) versus the power supply voltage and (b) the input sensitivity in* mV$_{rms}$.

3.3.4.2 Phase Noise Results

To measure the residual phase noise of the dual-modulus prescaler, the measurement setup of Fig. 3.11 (a) is used. A signal generator drives two devices under test (DUT), two prescalers in this case. One of the prescalers divides constantly by 8, while the other one is manually switched between 8 and 9 to obtain 90° shifted inputs for quadrature locking in the phase noise measurement system. The phase noise measurement system is the Europtest PN9000. The noise measured at the output of the phase detector consists only of the residual noise of the prescaler,

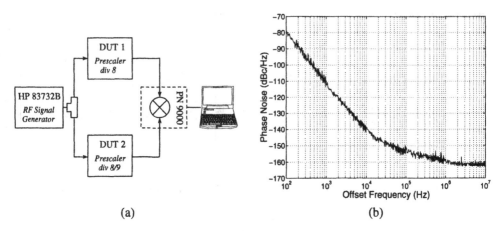

(a)   (b)

Figure 3.11: *Experimental results for the DMP in 0.7 μm CMOS: (a) Prescaler phase noise measurement setup and (b) measured phase noise performance.*

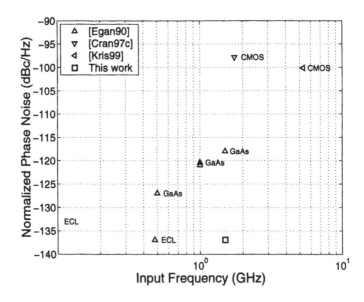

Figure 3.12: *Comparison of the phase noise performance of published prescalers and this work. The phase noise is taken at 100 kHz offset and normalized towards the input frequency of the prescaler.*

since the correlated noise of the reference source is cancelled by the mixing action. Assuming that the noise of both prescalers is independent and that the circuits are equal, the sum of the residual phase noises is measured, so the noise of one prescaler is 3 dB lower than the measured value.

Phase noise measurements are shown in Fig. 3.11 (b). As explained earlier, three regions can be distinguished, a $1/f^3$ region, a $1/f$ region and a noise floor. The measured output phase noise is -112 dBc/Hz at 1kHz, -152 dBc/Hz at 100 kHz and -163 dBc/Hz for the noise floor. The phase noise floor is close to the limits of the measurement setup due to the locked nature of a flipflop frequency divider. The phase noise spectrum is $1/f$ noise determined as predicted in [DeSm98]. In Fig. 3.12 the results are compared with other published prescalers. In order to compare the different prescalers, the output phase noise at 100 kHz offset is referred to the input using Eq. (3.2) and is plotted versus the maximum operating frequency. As mentioned earlier, the prescaler of [Cran98] is implemented in the same process but with CML-logic. The phase noise of the CML prescaler is almost 40 dB worse than of the presented prescaler, confirming the rules of the thumb in Section 3.3.3. To summarize, the presented prescaler offers robust operation with a small single-ended input, exhibiting superior phase noise performance while maintaining a comparable frequency-power ratio as inherently faster CML-logic prescalers.

3.3.4.3　Design Issue: The Quadrature Accuracy

The measurements of the dual-modulus prescaler revealed a possible flaw of the phase-switching topology while dividing by 9. In the output spectrum, spurious tones of around -30dBc were present at $k/4$ of the output frequency with $k = 1, 2, \ldots$. These spurious tones are disastrous in phase-locked loops. The /4 dependency leads one to suspect that the source of these spurious is the quadrature accuracy in the half-speed divide-by-2 flipflop. As can be seen in Fig. 3.6, $F4.I$ and $\overline{F4.I}$ are fed back over the output inverters, while the other outputs are not. This leads to an unwanted delay between the 90° shifted phases; As a result, every phase switch action causes a systematic error on the output edges, leading to the /4 spurious tones. To prove this hypothesis, a FIB (Focused Ion Beam) operation is performed to remove the inverters from the feedback loop. Simulations indicate that this greatly reduces the unwanted delay. In accordance, measurements show a decrease in the power of the spurious tones of at least 15 dB.

In short, to maintain proper operation of the phase-switching dual-modulus prescaler, the inherent perfect quadrature properties of the half-speed master/slave flipflop should be cherished. Moreover, in the layout of the circuit care must be taken to equalize the connection delays of all four phases to reduce the unwanted spurious tones to negligible levels.

3.4　A Single-ended 1.8 GHz 8/9 DMP in 0.8μm "Radiation Hardened" BiCMOS

3.4.1　The Circuit Implementation

To implement a high-speed dual-modulus prescaler in a 0.8μm BiCMOS technology, without using the bipolar transistors, all possible speed improving possibilities have been explored and implemented, within the constraint of power consumption. Therefore, the dual-modulus function

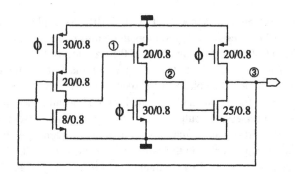

Figure 3.13: *The high-speed divide-by-2 single-ended D-flipflop of the 0.8μm BiCMOS dual-modulus prescaler.*

is realized with the phase-switching architecture. The speed bottleneck in the prescaler design is redirected to the high-speed D-flipflop.

The design target of the high-speed D-flipflop is an as high as possible speed with a small, single-ended input signal for a reasonable power consumption. To go to very high frequencies, the CML-like (Current Mode Logic) D-flipflop (Fig. 3.4 (b)) is an option for its very good speed-power ratio. However, the CML-like flipflop needs differential input signals, requiring a high-speed single-ended to differential converter. Such a circuit integrated in a 0.8μm technology consumes vast amounts of power. Moreover, due to its sensitivity to the DC input, the CML-like flipflop is not sufficiently robust for the given application. A more viable option are TSPC (True-Single-Phase-Clock) dynamic D-flipflops [Yuan89], which combine both single-ended and high-frequency operation. Unfortunately, the high-frequency operation is obtained with almost square-wave rail-to-rail input signals. The design challenge is to find a circuit that can operate with small input signals, without the input buffering that usually consumes more power than the prescaler itself [Chan96, Rog94].

The nine-transistor edge-triggered TSPC D-flipflop of Fig. 3.4 (a) is taken as a starting point. A first improvement is to remove the clock transistor from the critical path of the flipflop by placing them as close to AC ground as possible. This gives a speed head start of almost 20 %. Besides, the clock transistors can be widened to increase the input sensitivity, without affecting the speed. A second speed bottleneck are the stacked NAND-type structures in the flipflop of Fig. 3.4 (a), which represent a rather large RC delay. Omitting the stacked transistors leads to race situations and to a DC path from power to ground in certain states, seriously increasing the power consumption. However, for high-speed divide-by-2 operation, the DC path only exists for a short time and hardly affects the power. Additionally, all states and their transitions are well known, race situations can be predicted and countered by proper transistor sizing. Another advantage of omitting the stacked transistors is that the load for the preceding stage is decreased. As is the case in all high-frequency design, the rule "less is more" is again valid.

The final D-flipflop is shown in Fig. 3.13. From left to right, it consists of a non-pre-charged (static) p-stage, which is similar to a C^{2}MOS p-stage. The two following stages, one pre-charged p-stage and one static p-stage in a pseudo-NMOS like configuration form a p-latch. This topology has been chosen over the static p-stage, pre-charge n-stage, static n-stage topology of Fig. 3.4 (a), since it offers some speed improvement. The functionality is kept equal by playing with relative transistor sizing. Again, the transistor sizing is a compromise, since still four transistor are loads as well as drivers. If $\phi = 0$ and ③ $= 1$, the first stage is transparent and pulls ① down, causing the pre-charge of ②. The p-latch is in its evaluation state and the output must change from 1 to 0. Therefore, the pull-down strength of the NMOS (i.e. the W/L) in the last stage must be larger than the pull-up strength of the PMOS, to keep the output (③) low. Due to the fast state transitions, the DC power consumed by this ratio-ed stage is only small. The same transition speed prevents the first stage from unintentionally charging ① when the clock is still low and the output changes from high to low. The first stage is implemented with stacked transistors to keep ① from charging when the clock changes state. When $\phi \rightarrow 1$, the static p-stage and the p-latch are in hold mode, but the intermediate node ② changes from 1 to 0 to close the NMOS, such that the output can become one in the next state. Again, the operation depends on the relative W/L ratios of the transistors, i.e. the NMOS must be strong –large– enough to pull down ②; The voltage of ② must remain below the V_T of the output NMOS. In the next state, $\phi \rightarrow 0$ and the output is charged to 1, to prepare for the next state. For speed and input sensitivity the PMOS of the last stage is quite large, meaning that the NMOS must be even larger. This increases the power consumption but is necessary to drive the load capacitance presented by the connections and the next divide-by-2 stage. When the output goes high, ①, which was pre-charged to one, is again pulled down. The NMOS is made small to slow down the discharge of ① and thus the turn-on of the PMOS in the second stage. When the clock goes high, the NMOS of the second stage must keep ② down, to keep the p-latch in hold mode as two states ago. The transition between the two previous states is hazy and overlapping, but cause the earlier mentioned speed enhancement. As a result, the presented D-flipflop is certainly not a fully functional flipflop and its low frequency division operation is impaired. But it satisfies the design target, i.e. high-speed with a small, single-ended input.

The small input constraint seriously limits the attainable maximum operating frequency. Simulations show that for a small input signal the maximum operating speed of the stand-alone flipflop is 2.15 GHz for 2.8 mA at 5V. For a rail-to-rail sine wave input signal, the maximum operating speed is as high as 2.9 GHz for 3.6 mA at 5V. To prove the higher performance of the presented flipflop, the nine-transistor TSPC latch of Fig. 3.4 (a) is designed and simulated with the clock transistors placed as close to AC ground as possible. Simulations reveal an over 15% higher speed-power ratio for the presented flipflop.

Since the high-speed flipflop of Fig. 3.13 is fully single-ended, the half-speed divide-by-2 flipflop must be able to provide a quadrature output for the phase-select block with a single-ended input. DSTC (Dynamic-Single-Transistor-Clocked) latches [Yuan97] are employed as in Fig. 3.6 but with one DSTC n-latch and one DSTC p-latch to implement the master/slave flipflop. This enables single-clock operation with only two clock transistors, minimizing the load for the high-speed flipflop.

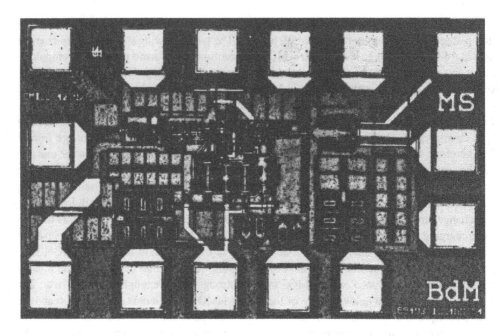

Figure 3.14: *IC microphotograph of the dual-modulus prescaler in 0.8μm BiCMOS.*

3.4.2 Experimental Results

An IC microphotograph of the 8/9 dual-modulus prescaler is shown in Fig. 3.14. The prescaler is integrated in a 0.8μm "radiation hardened" BiCMOS technology. The technology offers 1.2 μm NPN bipolar transistors, but a CMOS only implementation is chosen to prove the high-frequency capabilities of CMOS through smart circuit design. The IC is aimed for space applications where radiation induced latch-up is critical for the circuit operation. The technology is hardened to tolerate radiation of more than 10Mrad and 10^{14} neutrons/cm^2 and of photons. The combination of a SIMOX buried oxide and trenches eliminate latch-up from charge accumulation by radiation.

The input signal is attenuated by the input bonding wire and the bonding pad capacitance. To boost the incoming signal the on-chip input is terminated with a 200Ω resistance, which causes an overshoot in the input transfer function in comparison with a 50Ω termination.

The measured maximum operating frequency of the prescaler is 1.8 GHz at a power supply voltage of 5V. In Fig. 3.15 (a), the maximum operating frequency and the corresponding power consumption (dashed) is plotted versus the power supply voltage. All measurements are performed at room temperature for small signal operation (< -5dBm) at a fixed DC input. The power consumption at 5V is 8.5 mA for divide-by-8 operation. For comparison with the 0.7μm CMOS implementation, the frequency-power ratio at 3V is calculated to be 95 MHz/mW. Although the technology is inherently slower, the ratio is even better than the 0.7μm CML implementation due to the speed-power optimization of the high-speed flipflop.

The power consumption of the prescaler is spread over the building blocks as follows (simu-

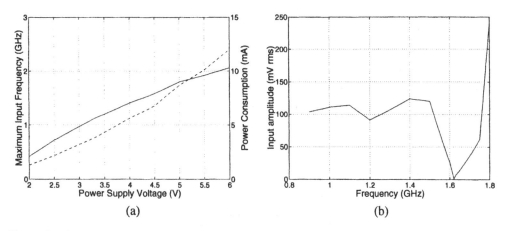

Figure 3.15: *Experimental results for the DMP in 0.8 μm BiCMOS: (a) The maximum input frequency and power consumption (dashed) versus the power supply voltage and (b) the input sensitivity in* mV_{rms}.

lation): High-speed D-flipflop: 2.7 mA, Half-speed D-flipflop: 2.4 mA, phase-select: 1.5/2.5mA, low-speed D-flipflop: 0.9 mA. The total simulated current is 7.5 mA, which is 15% lower than measured. Again, the inconsistencies are attributed to the poor high-frequency modeling of the transistors. However, the simulated and measured maximum operating frequency of the full prescaler are in close correspondence and determined by the speed of the half-speed divide-by-2 flipflop.

In Fig. 3.15 (b), the input sensitivity is plotted. The maximum input sensitivity over the whole frequency range is below $250mV_{rms}$, taking into account the attenuation presented by the package parasitics. This is again orders of magnitude smaller than other presented CMOS prescaler [Chan96, Rog94], without using input buffers. This proves that the realization of high-speed prescalers with a small, single-ended input signal is feasible even in "ancient" CMOS technologies.

3.5 A 1.8 GHz 16-modulus /64-/79 Prescaler in 0.25μm CMOS

3.5.1 The Divide-by-2 Flipflops

The 1.8 GHz 16-modulus prescaler is intended for integration in a fully integrated phase-locked loop frequency synthesizer for DCS-1800. Therefore, it can benefit from the large, differential signal at the VCO output to optimize the division operation towards power consumption. The residual phase noise is of less importance, since the in-band phase noise of the PLL is mainly determined by the passive loop filter components. Nevertheless, a dynamic CMOS implementation is chosen for low residual phase noise and for robustness. The required specifications of the DCS-1800 prescaler, derived from Table 2.2, are summarized in Table 3.1. For the high-speed

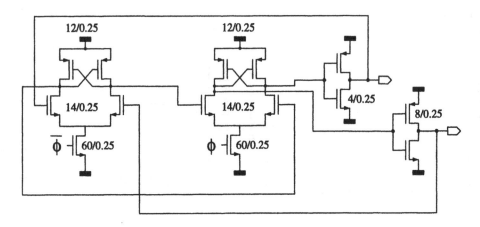

Figure 3.16: *The high-speed divide-by-2 differential D-flipflop of the 0.25µm CMOS multi-modulus prescaler.*

Center Frequency	1.8 GHz	Section 2.6.2
Tuning range	170 MHz	Section 2.6.2
Power Consumption	*ALAP*	Section 2.6.1
Area	*ALAP*	Section 2.6.1
Supply Voltage	1.8V-2V	Section 2.6.1

Table 3.1: *Summary of the frequency synthesizer specifications in a DCS-1800 communication system.*

division, a CML-like flipflop (Fig. 3.4 (b)) presents a better speed-power trade-off, but the possible power advantage is alleviated by the need of amplifiers to convert the small divider output signals to CMOS logic levels for the phase-select block.

For the implementation of the high-speed divide-by-2 flipflop the double n-latch DSTC D-flipflop of Fig. 3.16 has been chosen. The design target is robust operation for a sub-2V power supply (Table 3.1). The transistor sizing is shown in Fig. 3.16 and is such to maintain a large internal signal swing, even at low operating voltages. The inverters buffer the second n-latch from the capacitive load presented by the half-speed D-flipflop; In that way, the frequency performance is maintained for loaded conditions. The clock transistors are rather large since the DC output of the VCO buffers and thus the $V_{GS} - V_T$ of the clock transistors is rather low (< 0.5V, see Chapter 4).

The half-speed divide-by-2 flipflop is similar to the high-speed one. The design is focused on presenting an accurate quadrature signal to the phase-select block to minimize the spurious tone generation. The inverters in Fig. 3.16 are omitted to exploit the inherent perfect quadrature relationship of the signals in a double n-latch master/slave flipflop. Also the layout is optimized towards equal delays in the quadrature path. The clock transistors are smaller ($W = 15µm$) than in the high-speed case, since the high-speed D-flipflop DC output is high enough. The sizes of

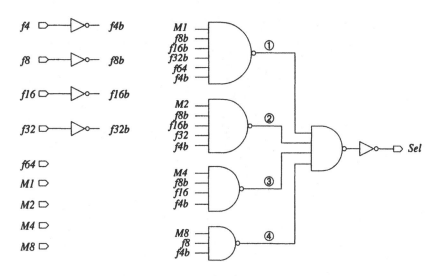

Figure 3.17: *The 16-modulus mode logic.*

the other transistors are approximately halved and designed to drive the load of the phase-select block.

3.5.2   The Multi-Modulus Implementation

The frequency synthesizer, in which the prescaler will be embedded, must be able to synthesize the full frequency range of DCS-1800, since the synthesizer provides the local oscillator signal for the receive and the transmit path. As stated in Chapter 2, the frequency range is from 1710 to 1880 MHz (Table 3.1). The reference frequency of the synthesizer is 26 MHz, meaning that the divider must be able to divide by division moduli from 65 to 73. To accommodate the modulus range, a 16-modulus prescaler dividing by 64 to 79 is implemented. The topology is the same as in Fig. 3.2 (a), except for the mode logic. The 16-modulus mode logic is shown in Fig. 3.17. The signals are named according to the division, i.e. $f4$ is the input divided by 4, ... When $M1$ is set high, the upper 6-input NAND outputs (①) a pulse with the width of half a period of $f4b$ every time all input signals are high, i.e. every output period. This means that a division by 65 is performed. Similarly when $M2$ is high, two pulses are generated every output period, i.e. a division by 66. For $M4$ four pulses are generated and for $M8$ 8 pulses, leading to division by 68 and 72. By combining the pulses in the output NAND gate, all division moduli between 64 and 79 can be realized. The different division signals $f4 - f64$ are delayed relative to each other since the low-speed division block is asynchronous; The mode logic is made insensitive to the relative delay by smartly combining the different division signals. All division signals are used to generate the control pulses, such that 16 is the maximum number of moduli attainable with this topology. To extend the modulus range, a frequency divider with one prescaler and two digital counters (Fig. 2.16) needs to be designed.

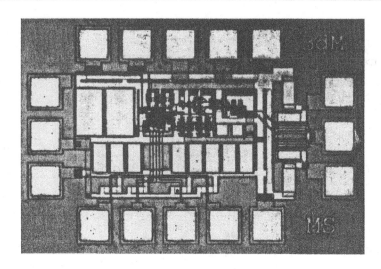

Figure 3.18: *IC microphotograph of the 16-modulus prescaler in 0.25μm CMOS*

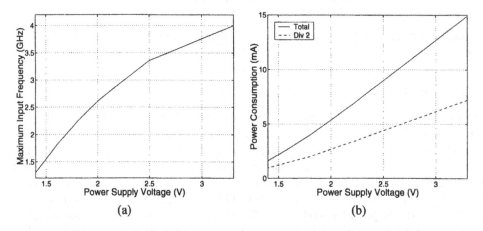

Figure 3.19: *Experimental results for the 16-modulus prescaler in 0.25 μm CMOS: (a) The maximum input frequency versus the power supply voltage and (b) the power consumption of the total prescaler and of the high-speed divide-by-2 flipflop versus the power supply voltage.*

3.5.3 Experimental Results

An IC microphotograph of the stand-alone 1.8 GHz 16-modulus prescaler is shown in Fig. 3.18. The four inputs to control the division modulus are clearly visible in the lower part of the figure.

In Fig. 3.19, the maximum input frequency and the power consumption are plotted versus the power supply voltage. The power consumption is measured at the maximum operating speed for the given supply voltage. At 2V, the maximum frequency is 2.61 GHz for a power consumption of

5.4 mA, revealing a speed-power ratio of 240. At 1.8V, those values are 2.24 GHz and 4 mA. The typical operation mode of the prescaler is between 2 and 1.8V for an input frequency of 1.8 GHz; The power consumption is then 10 and 7 mW respectively. The presented D-flipflop is capable of 4 GHz operation at 3.3 V. As can be seen in Fig. 3.19 (b), the power consumption of the high-speed D-flipflop is approximately half of the total power consumption of the prescaler for divide-by-64 operation. The simulated power consumption is spread over the different building blocks as follows: High-speed flipflop: 2.2mA, Half-speed flipflop: 0.85mA, phase select: 1.2mA (/64)-2.4mA(/79), low-speed flipflops: 0.5mA. The simulated power consumption of 9.5mW at 2V and 1.8GHz is in close correspondence with the measured values (within 5%), since HSpice Bsim3v3 level 49 transistor models are employed which incorporate high-frequency effects.

The spectral purity measurements of the PLL (Chapter 5) show no spurious tones whatsoever for none of the division moduli. This proves that careful design and layout to maintain the quadrature accuracy is effective to reduce the spurious tone generation to negligible levels.

3.6   A 12 GHz /128 Prescaler in 0.25μm CMOS

3.6.1   Introduction

The ever-growing demand of the industry for smaller and cheaper solutions has pushed CMOS to be the most promising technology for fully integrated complex systems with dense logic, low power and low power supply voltage. However, in very high-speed applications (around 10 GHz) the use of CMOS is impeded by its relatively low transconductance as opposed to silicon bipolar or SiGe devices. This imposes a severe power-speed trade-off. Fortunately, the down-scaling of CMOS device sizes relaxes this trade-off, making deep sub-micron CMOS an attractive solution for multi-GHz applications. As an example, a 12 GHz divide-by-128 frequency divider is integrated in a first-generation 0.25μm CMOS technology. The gate oxide thickness is 5.8 nm for the NMOS transistor and 6.6 nm for the PMOS transistor. The minimum gate electrode length is 0.25μm, but the overall design rule is that of a 0.6μm process. As a result, the parasitic capacitance present at the transistor terminals will be substantial, making the technology less suitable for high-speed applications. To show that the technology imposes serious limits to the divider performance, the oscillation frequency of a three-stage, minimum-sized ring oscillator is simulated for different 0.25μm CMOS technologies. A three-stage ring oscillator is chosen, because dynamic CMOS flipflops (see Fig. 3.4 (a)) are in fact ring oscillators which are forced to operate at a certain frequency by an external clock. Its operating frequency is an upper limit for the maximum internal flipflop frequency. The maximum frequency that can be divided is then twice the internal flipflop frequency.

The 0.25μm technology used for this design (This tech) is compared to three other 0.25μm technologies in Fig. 3.20. It can be seen that the 0.25μm technology employed here is outperformed by the others. The maximum flipflop divide-by-two frequency is around 4.45 GHz at 2 V and 10.2 GHz at 4 V for this technology. The performance of the more advanced 0.25μm CMOS technologies (Tech1) is more than twice as good, i.e. 9.16 GHz at 2 V and 15.95 GHz at 4 V.

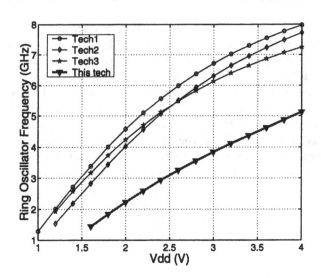

Figure 3.20: *Oscillation frequency of a minimum-sized three-stage ring oscillator for different 0.25μm CMOS technologies versus power supply.*

3.6.2 The Circuit Implementation

3.6.2.1 The High-Speed Divide-by-2 Flipflop

The circuit of the high-speed divide-by-two flipflop is given in Fig. 3.21. The master and slave latches are clocked by differential clock signals. Both latches are composed of a clocked differential sensing amplifier pair and an inversely clocked latch pair. The key to the high-speed operation of the flipflop is the limitation of the internal swing. The smaller the transitions that have to be made, the faster the latches can change state. The simulated internal peak-to-peak voltage is 0.6 V at 12 GHz. Only NMOS transistor are implemented to limit the parasitic drain capacitance and to minimize the power consumption. Compared to the standard CML-like latch in Fig. 3.4 (b), the biasing current sources are omitted, resulting in a speed increase of around 10%. The DC bias point of the circuit in Fig. 3.21 is determined by the size and $V_{GS} - V_T$ of the clock transistors, i.e. the DC level of the clock input, and the value of the load resistance.

A first constraint is the pole formed by the resistors and the drain capacitances, the parasitic capacitances of the resistors and interconnection and the load of the next stage. To make this pole high enough ($\approx 9GHz$), the resistance must be small (300 Ω), which leads to increased power consumption to set the DC output. This DC output level determines the DC bias point of the next flipflop; To be useful in a system, the high-speed flipflop must not only be optimized for high speed but also for providing the proper DC output level. This is crucial for the operation of the next dividers, since high-speed CML-like dividers are very sensitive to the DC input level.

The second constraint is set by the input sensitivity of the flipflop. Since the goal is to supply the input signal without RF probes nor flip-chip bonds, the on-chip input signal will

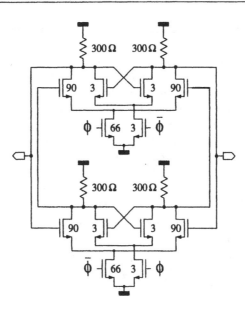

Figure 3.21: *The 12 GHz divide-by-2 CML-like D-flipflop circuit. All sizes are in μm (if not mentioned otherwise) and all transistors gates are minimum length, i.e. 0.25μm.*

be significantly attenuated with respect to the off-chip signal. To make sure that the clock can synchronize the flipflop, the clock transistor transconductance must be large (16 mS). This results in increased power consumption and an increased sensitivity to the DC clock level, since the small signal must still be able to drive the transistor from the linear region to saturation. Another trade-off is present in the transistors of the latch pair. When their transconductance is large, fast latching is possible, but changing state will be more difficult. Because the clock is not a square rail-to-rail wave, but a small sinusoidal signal, a grey area is present between the sensing and the latching. When the latching is strong, the sensing will have to be strong too, leading to an increased power consumption. As a result, the latching transistors in the high-speed flipflop (see Fig. 3.21) are small ($W = 3\mu$m) compared to the size of the sensing differential pair.

3.6.2.2 The Divide-by-128 Prescaler

The total frequency divider is composed of 4 high-speed divide-by-two flipflops, similar to the one in Fig. 3.21, and 3 TSPC dynamic flipflops [Yuan89], which provide the lower frequency division with low power. The sizes of the clock and sensing transistors of the succeeding high-speed flipflops are approximately halved when the frequency is halved. To provide a higher signal swing, the width of the latching transistors is increased to 12μm. An important issue in the design of cascaded flipflops of the type of Fig. 3.21 is the clock DC level of one flipflop, which is the DC output level of the preceding flipflop. As mentioned earlier the division operation is highly sensitive to the DC levels. Also the amplitude of the output, which serves as clock signal must be sufficiently high to ensure proper operation of the next flipflop. Most published dividers

[Raz95, Wang00] only mention the high-speed divide-by-two flipflops, but to operate at system level, the flipflops need to be designed to drive the next stages. This imposes a limit on the maximum operating frequency since the output amplitude of the flipflop must be high enough and the DC biasing point will not be the optimal bias point for high-speed operation. A solution is to insert level shifters between two successive flipflops. However, the signal amplitude will be attenuated. Due to the body effect, the gain of the level shifter at frequencies lower than the pole determined by the load capacitance and the transconductance of the level shifter transistor, is less than unity. At frequencies above this pole, the gain drops until it is determined by the capacitive division by the load capacitance and the gate-source capacitance. The resulting signal amplitude at high frequencies is too small to drive the next divider. A capacitive AC coupling can be inserted with a resistive division to determine the DC input level of the next flipflop. Due to the parasitic capacitance to ground of the coupling capacitor, the same capacitive division is present as before or the load for the first flipflop becomes too large.

To prevent signal attenuation, the successive dividers are directly connected. To ensure proper division over a large frequency range, all flipflops have to be designed as a single block. For lower power consumption, the second flipflop is designed at a low power supply voltage. Therefore, the DC output of the first flipflop can not be too high to accommodate the second flipflop. Three possible solutions exist. The power supply of the first flipflop can be lowered or the load resistor can be increased, but both actions deteriorate the frequency performance. The other solution is to increase the current in the first flipflop, which also leads to a higher input sensitivity. Due to the fact that the output amplitude of the first flipflop must be high enough to be useful as a clock signal, the maximum operation frequency of the total divide-by-128 will be lower than that of the stand-alone high-speed flipflop.

3.6.2.3 The Input Section

The measurements have been performed without using RF probes nor flip-chip bonding. The IC is glued on an aluminum oxide substrate. The connection between the 50Ω micro-strip lines on the substrate and the IC is done with gold bondwires (see Fig. 3.22 (b)). As a result, the input signal will be severely attenuated.

The on-chip input section consists of an on-chip termination resistor, together with a coupling capacitance and a DC bias resistor. The termination resistor is chosen larger than 1000 Ω to increase the on-chip signal. A second degree of freedom is the bondwires parasitic inductance. To design the bondwire inductance, the following formulas are used [Gree74]:

$$L = \frac{l}{5} \cdot \left[\ln\left(\frac{2l}{r}\right) - 0.75 + \frac{r}{l} \right] - F \tag{3.4}$$

$$M = \frac{l}{5} \cdot \left[\ln\left(\frac{l}{d} + \sqrt{1 + \left(\frac{l}{d}\right)^2}\right) - \sqrt{1 + \left(\frac{d}{l}\right)^2} + \frac{d}{l} \right] \tag{3.5}$$

with L the inductance, M the mutual inductance, l the bondwire length, d the distance between 2 bondwires, r the radius of the bondwire and F a factor that takes the frequency dependence of

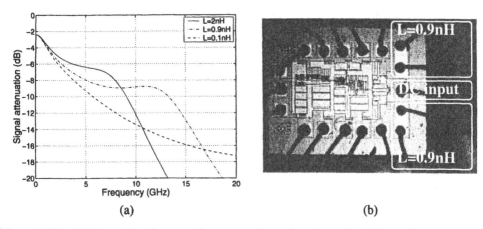

Figure 3.22: *(a) Input signal attenuation versus input frequency for different bondwire inductance values. (b) A microphotograph of the input bondwires.*

the inductance into account. To obtain an inductance of 0.9 nH with 2 bondwires and knowing that $r = 12.5\mu$m, the length and distance can be calculated to be $l = 1.2$ mm and $d = 0.6$ mm.

The resulting inductance is used to boost the input signal in the desired frequency range. To choose the required bondwire inductance, a model of the input section has been developed that takes into account all on- and off-chip parasitics. Fig. 3.22 (a) shows the simulated transfer function of the model for different values of the bondwire inductance. A value of 0.9 nH is chosen for the frequency range around 12 GHz. As can be seen in Fig. 3.22 (a), in case of flip-chip bonding ($\approx$ 0nH) the signal attenuation would at least 5 dB higher.

To make a differential input a delay is inserted in one of the inputs. The delay was calibrated using a 6 GHz network analyzer by realizing a 45° shift at one fourth of the input frequency. Due to bonding wire matching constraints and the limited calibration accuracy, the input signal to the circuit is not perfectly differential. This is however not critical, since CML-like circuits can operate with near-differential inputs, i.e. with an error of less than 10°.

3.6.3 Experimental Results

Fig. 3.23 shows a microphotograph of the circuit. The active area is 500 x 200 μm ^{2}. Fig. 3.24 (a) shows the measured maximum input frequency versus the power supply voltage for the high-speed divide-by-2 flipflop. At 2V, the maximum input frequency for divide-by-128 operation is 12 GHz. Fig. 3.24 (b) presents the speed-power trade-off. As a comparison, the dividers in [Raz95, Kur97], which are implemented in advanced 0.1μm and 0.15μm CMOS, achieve at 2 V only 10 GHz and 11.8 GHz (see Table 3.3). The presented high-speed flipflop can divide up to 12 GHz consuming less than 20mW, which is better then the advanced 0.1μm /0.15μm CMOS implementations of [Raz95, Kur97]. The input sensitivity of the first divider is plotted in Fig. 3.25 (a), taking into account the input attenuation. The sensitivity measurements are performed under fixed input conditions. For higher frequencies, different biasing conditions are

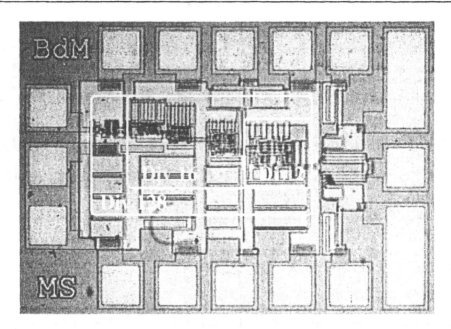

Figure 3.23: *An IC microphotograph of the 12 GHz /128 prescaler in 0.25μm CMOS.*

	Range	Vdd	Power (at 12 GHz)
Divide-by-2	3-15 GHz	2V	20mW
Divide-by-16	9-14 GHz	2.4/2V	56mW
Divide-by-128	10-12 GHz	2.4/2V	60mW

Table 3.2: *Summary of measured performance of the different divider stages.*

applied such that the lowest input signal is around -3 dBm. Compared to [Raz95, Kur97], the input sensitivity is an order of magnitude lower. Fig. 3.25 (b) shows the output waveform (93.75 MHz) for a 12GHz input signal. The measured power consumption is 32mW at 2.4V for the high-speed divider and 28mW at 2V for the remaining dividers. Under the same conditions, the frequency divider divides by 128 in a frequency range between 10 and 12 GHz. As mentioned earlier, the frequency performance of the total divide-by-128 circuit is worse than that of the stand-alone flipflop; High-speed divide-by-16 operation is obtained in a range from 9 to 14 GHz (see Table 3.2). The DC bias issues are the reason for the limited maximum operating frequency and frequency range as compared to the stand-alone divider, which explains the high operating frequencies of [Raz95, Wang00]. The stand-alone high-speed flipflop of this design operates with input frequencies up to 15 GHz. Table 3.3 provides an overview of previously published high-speed dividers. It can be seen that the presented high-speed divider combines high speed, high input sensitivity and multi-stage division, while integrated in first-generation 0.25μm CMOS.

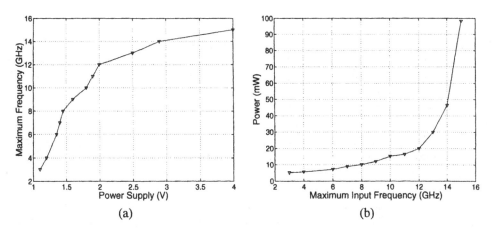

Figure 3.24: *Experimental results for the 12 GHz prescaler : (a) Maximum input frequency versus the power supply voltage. (b) Measured power consumption of the high-speed flipflop versus the maximum input frequency.*

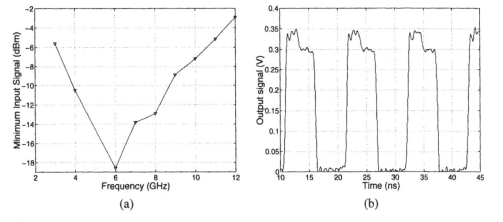

Figure 3.25: *Experimental results for the 12 GHz prescaler : (a) Minimum input signal versus the input frequency. (b) The output waveform for a 12 GHz input signal divided by 128.*

	Freq (GHz)	/N	Power (mW)	Tech (μm)	Input
[Raz95]	13.4	/2	28	0.1	2.6V
[Kur97]	11.8	/64	31	0.15	11 dBm
[Wang00]	16.8	/2	3	0.25	-3dBm
This work	12	/128	60	0.25	-3dBm

Table 3.3: *A comparison of different published high-speed CMOS dividers.*

3.7 Conclusion

This chapter has covered the design and implementation of high-speed prescalers. The additional logic needed to implement the dual-modulus division seriously degrades the speed performance of the prescaler as opposed to a stand-alone flipflop. Moreover, the flipflops need to be fully functional while a stand-alone flipflop can be optimized for divide-by-2 operation. The total speed degradation can be up to a factor 2! Therefore, the phase-switching architecture is adopted to implement the dual-modulus function; This allows full exploitation of the inherent speed of a stand-alone flipflop by switching between quadrature phases of the divide-by-4 signal, generated by a master/slave flipflop. The 90° phase relationship between the signals needs to be as accurate as possible to prevent the generation of spurious tones in the prescaler output at $k/4$ of the output frequency with $k = 1, 2, \ldots$.

The prescaler design comes down to optimizing the trade-off between speed and power in the high-speed flipflops. Fig. 3.26 summarizes the prescalers presented in this book. The additional features of the different prescalers are listed to emphasize the fact that the prescalers are not only optimized towards speed and power but also to operation with small, single-ended input signals, low residual phase noise, robust operation or extremely high speed respectively. The phase noise generation mechanisms in prescalers are discussed in this chapter and rules of the thumb are given to minimize the residual phase noise.

The main trade-off in prescaler design is between speed and power; The prescaler is actually a high-speed digital circuit, implying that its operation follows the dynamic power law of digital circuits.To study the effect of scaling, the speed-power ratio of the presented prescalers is plotted together with that of published CMOS dual-modulus prescalers in Fig. 3.26. The lines in the figure indicate the theoretical speed-power ratio, i.e. $f/P \propto (C_L \cdot V_{DD}^2)^{-1}$. Typical V_{DD} values for the given technology are chosen in correspondence to the values in the publications as shown in the figure. The load capacitance C_L for a given technology is computed from available HSpice transistor models for an inverter with the PMOS twice as large as the NMOS. The relevance of the trend lines is mainly qualitative, but shows clearly that high-speed prescaler design greatly benefits from the scaling of CMOS technology and scaling of the power supply voltage.

As can be seen in Fig. 3.26, the performance of the dual-modulus prescaler in 0.8μm BiC-MOS is better than the predicted trend, due to the dedicated high-speed flipflop design; The speed-power ratio is as high as 95 MHz/mW at 3V. The 0.7μm CMOS prescaler is slightly worse than the predictions, since it is not only optimized for speed; For low phase noise a differential implementation is chosen that almost doubles the power consumption. The speed-power ratio of 78 MHz/mW at 3V is however comparable to a CML-implementation in the same technology [Cran98]. The performance of the 0.25μm CMOS 16-modulus prescaler is close to the power law, although the logic for the 16-modulus division increases the power consumption. The lowest ∇ at 0.25μm is the CML-like 12 GHz /128 prescaler; The lower ratio is attributed to the input DC sensitivity of this type of circuits, which impede the prescaler to operate at the full speed of a stand-alone flipflop (highest ∇).

The implementations and issues discussed in this chapter prove the feasibility of fully integrated, high-speed CMOS prescaler design for different applications and specifications.

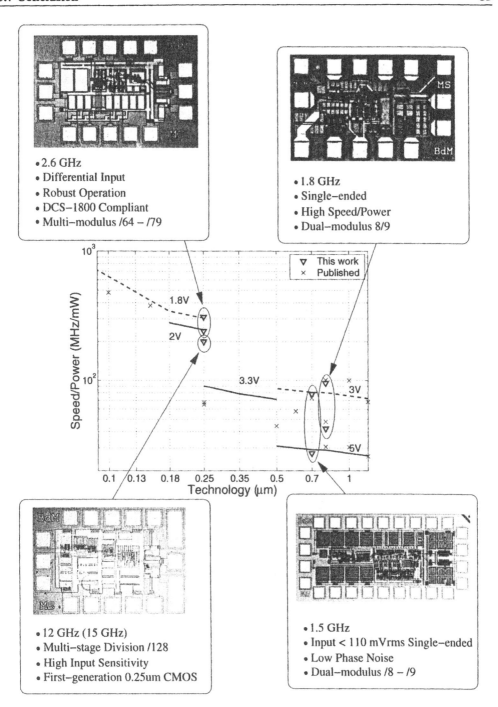

Figure 3.26: *The theoretical effect of CMOS scaling on prescaler performance compared to published state-of-the-art CMOS prescalers and the prescaler presented in this work.*

Chapter 4

Monolithic CMOS LC-VCOs

4.1 Introduction

Together with the prescaler, the Voltage Controlled Oscillator (VCO) operates at the highest frequency in a frequency synthesizer. The VCO is however more critical; The phase noise of the VCO determines the out-of-band noise of the synthesizer. For high-quality communications, passive LC-tank oscillators tend to be a better choice than relaxation/ring oscillators or active inductors to combine high frequency and low noise under power constraints (see Chapter 2).

Due to the importance of VCO noise in communication systems, a design-oriented phase noise theory is presented that takes into account the non-linearity of the active element. The presented theory provides a transparent expansion of linear theory, that enables the designer to quickly evaluate the influence of the circuit design on the overall phase noise performance by definition of the excess loop gain, starting from calculations presented in [Boon89].

The quality of the integrated LC-tank of the VCO is of utmost importance for the phase noise. Unfortunately, in CMOS the quality of the inductor is limited by metal losses (DC, skin effect, eddy current) and substrate losses. Since these effects are too complex to analyze analytically, a simulator-optimizer program has been developed that provides an optimal coil geometry for a given circuit topology and specification set in a given technology. Inductor design guidelines are given together with different varactor implementations in standard CMOS.

Next, VCO circuit design is discussed; CMOS has been recognized as a more powerful technology than bipolar for high-frequency, low-phase-noise VCOs. The most power-efficient VCO topology has been discussed, depending on the design conditions and technology. Finally, $1/f^3$ phase noise mechanisms are elaborated, a method to measure the sensitivity of a VCO circuit to 1/f noise upconversion is presented and a bias filtering technique is experimentally verified.

All the prior theory is applied; First, a 2 GHz VCO set with flicker noise upconversion minimization is presented with phase noise as low as -125 dBc/Hz at 600kHz. In a small band, the phase noise is as low as -132.5 dBc/Hz at 600kHz with no flicker noise at all, due to current source filtering. Secondly, a fully integrated VCO with a tuning range of 28% and a phase noise of less than -127.5 dBc/Hz at 600kHz is discussed, which exceeds the severe DCS-1800 specification using a 2-metal, standard 0.25μm CMOS technology. Additionally, the VCO is designed to drive a poly-phase filter for accurate quadrature outputs in a power-optimized manner.

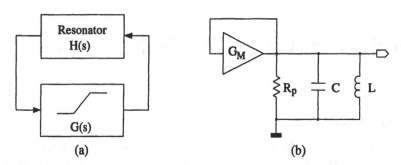

Figure 4.1: *(a) The basic oscillator feedback network. (b) The basic LC-tank oscillator schematic.*

4.2 General Oscillator Theory

An LC-tank oscillator basically is a feedback network consisting of a resonator $H(s)$ and a weakly non-linear active element $G(s)$ (see Fig. 4.1 (a)). If the oscillator has a resonator filter of at least an order of two, a sinusoidal output results, since the light harmonic content due to the weakly non-linear amplification is filtered by the resonator. A second order resonator can be realized by a series and a parallel connection of inductances and capacitances, but here only parallel connections are considered with a single parallel tank resistance R_p. The parallel resistance incorporates all series resistances of the passive elements, as shown in Eq. (4.1) in conductances with $G_{p,i}$ the intrinsic parallel conductance in the tank, $G_{s,L}$ and $G_{s,C}$ the series resistances of the inductor and the capacitor respectively and ω_0 the oscillation frequency.

$$G_p = G_{p,i} + \frac{1}{(\omega_0 L)^2 G_{s,L}} + \frac{(\omega_0 C)^2}{G_{s,C}} \qquad (4.1)$$

The active circuit used to close the loop, must have a memoryless or time-invariant limiting transfer. The time-invariance allows more easy calculation and ensures that there is no phase shift in the active part such that the oscillation frequency is accurately set by the passive filter. The small signal gain of the active circuitry guarantees the small-signal instability and its combination with the limiting characteristic enables stable oscillation. A typical non-linear active transfer is depicted in Fig. 4.1 (a).

The condition for stable oscillation is satisfied at the frequency where the loop transfer function is equal to one: $G(s)H(s) - 1 = 0$. The complex oscillation condition can be reduced to two equations with real variables, i.e. the well-known Barkhausen criterion for oscillation [Bark35]:

$$\angle G(j\omega_0) + \angle H(j\omega_0) = 0 \qquad \text{"phase balance"} \qquad (4.2)$$
$$|G(j\omega_0)| \cdot |H(j\omega_0)| = 1 \qquad \text{"amplitude balance"}$$

In fact, the *phase balance* condition states that the oscillator signal goes around the loop and returns *in phase*, where it is added to the original signal, ensuring instability for oscillation. From this condition, the frequency of the oscillator can be calculated; Since the active element is weakly non-linear it can be approximated by its first harmonic response, i.e. its small signal gain G_M in the case of the oscillator in Fig. 4.1 (b). The loop gain is

$$GH(s) = G_M \cdot \frac{sL}{1 + s\frac{L}{R_p} + s^2 LC} \tag{4.3}$$

The imaginary part of the loop transfer function is then

$$\angle GH(s) = G_M \cdot \frac{\omega L \cdot (1 - \omega^2 LC)}{(1 - \omega^2 LC)^2 + \omega^2 \cdot \left(\frac{L}{R_p}\right)^2} = 0 \quad \Rightarrow \quad \omega_0 = \frac{1}{\sqrt{LC}} \tag{4.4}$$

The *amplitude balance* states that the loop gain must be (at least) one to ensure that oscillation is maintained. For the oscillator under discussion this condition translates to:

$$|G(j\omega_0)| \cdot |H(j\omega_0)| = G_M \cdot R_p = 1 \tag{4.5}$$

meaning that the active element must compensate for the losses in the tank resistance.

The transfer function of the resonator can be rewritten in the following form:

$$H(\omega) = \frac{sL}{1 + s\frac{L}{R_p} + s^2 LC} = \frac{H_0}{1 + jQv} \tag{4.6}$$

with $H_0 = H(\omega_0) = R_p$, Q the quality factor of the tank and v the "detuning" from ω_0 : $v = \frac{\omega}{\omega_0} - \frac{\omega_0}{\omega} \approx \frac{2\Delta\omega}{\omega_0}$ with $\Delta\omega$ a small offset from the carrier frequency ω_0 .

The quality factor Q of a (damped) LC-tank can be defined in different ways; For a typical parallel RLC circuit, the Q is defined as the ratio of the center frequency and the two-sided -3dB bandwidth $2B$, i.e. $Q = \omega_0/2B$. If the RLC is incorporated in a feedback network for an oscillator as in Fig. 4.1 (a) and looking at the phase of $GH(s)$, the Q can be calculated to be $1/2 \cdot \omega_0 \cdot d\phi/d\omega$ with $d\phi/d\omega$ the slope of the phase of the transfer function for a frequency change. Knowing the first Barkhausen condition (Eq. (4.4)), this definition shows that the Q is a measure for how much the oscillator opposes to variation in the frequency of oscillation, i.e. the frequency stability. The higher the Q the higher the frequency stability. A third, general definition of the quality factor is:

$$Q = 2\pi \cdot \frac{\text{Energy stored}}{\text{Energy dissipated per cycle}} \tag{4.7}$$

Using this formula, the quality of the LC-tank in Fig. 4.1 (b) can be calculated. Assuming that a sinusoidal signal $V(t) = V_A \cdot \sin(\omega t)$ is applied, the energy stored in the capacitor is maximum and the energy in the inductor zero, when the voltage over the capacitor is maximum (V_A). The stored energy is thus $E_{peak} = C \cdot V_A^2/2$. Energy is lost in the circuit in R_p:

$$E_{loss} = \int_0^{2\pi/\omega} \frac{[V_A \cdot \sin(\omega t)]^2}{R_p} dt = \frac{\pi \cdot V_A^2}{\omega \cdot R_p} \tag{4.8}$$

The quality factor of the LC-tank at resonance is than:

$$Q = R_p \cdot \omega_0 C = \frac{R_p}{\omega_0 L} = R_p \cdot \sqrt{\frac{C}{L}} \tag{4.9}$$

4.3   A Design-Oriented Non-Linear Phase Noise Theory

In this section, a design-oriented non-linear phase noise theory is presented that not only takes into account the noise of the passive and active devices, but also the influence of the non-linear active element in the oscillator, based on the calculations in [Boon89]. A linear phase noise analysis as in [Cran98, Lees66, Herz00] is straightforward, but disregards upconversion of noise at harmonics of the oscillator, due to the non-linear active element. These calculations therefore underestimate the phase noise with at least 3dB.

4.3.1   The Theory

The losses in the resonator and in the active components are afflicted with noise. This noise is transferred to the oscillator output where it manifests itself as parasitic phase and amplitude modulation. In the following calculations the noise is assumed to be the result of stationary Gaussian random processes and sufficiently smaller than the sinusoidal oscillator signal to enable small-signal transfer calculus.

4.3.1.1   Modeling of the Non-linear Active Element

Since the small-signal transfer of the resonator is straightforward, the small-signal transfer of the active element is analyzed. The non-linear active element is assumed to have a memoryless, i.e. time-invariant transfer characteristic $g(x)$. With at the input of the active element, a large sinusoidal signal $V(t) = V_A \cdot \sin(\omega t)$ and noise $e_{ni}(t)$, the output can be written by the first two terms of a Taylor expansion:

$$g\left(V_A \cdot \sin(\omega t) + e_{ni}(t)\right) = g\left(V_A \cdot \sin(\omega t)\right) + \left.\frac{dg}{dx}\right|_{x=V_A \cdot \sin(\omega t)} \cdot e_{ni}(t) \qquad (4.10)$$

This means that the output consists of the harmonics of the sinusoidal signal ($g(x)$ is weakly non-linear) and of the transferred noise. The small-signal or noise transfer function $\frac{dg}{dx}$ is –in contrast to the transfer characteristic $g(x)$– time-dependent, due to its dependency on the large input signal. Since the large input signal is periodic, the small-signal or noise transfer function $\frac{dg}{dx}$ is periodic and can be expressed using Fourier series with coefficients g_n.

$$e_{no,g}(t) = e_{ni}(t) \sum_n g_n e^{jn\omega_0 t} \qquad (4.11)$$

$$E_{no,g}(\omega) = \sum_n g_n E_{ni}(\omega - n\omega_0) \qquad (4.12)$$

with $E_{no,g}(f)$ and $E_{ni}(t)$ the Fourier transforms of $e_{ni}(t)$ and $e_{no,g}(t)$ respectively. In other words, the output noise in the frequency domain is the result of convolution of the input noise with a series of Dirac impulses at multiples of f_0, that describe the time-dependent noise transfer function $\frac{dg}{dx}$. The power spectral density of the output noise is then:

$$S_{no,g}(\omega) = \sum_n |g_n|^2 S_{ni}(\omega - n\omega_0) \qquad (4.13)$$

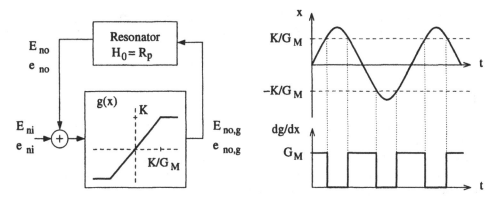

Figure 4.2: *The small-signal transfer of the active element and its dependence on the large sinusoidal oscillator signal.*

The noise power around all harmonics of the oscillator signal is converted to noise around the wanted carrier. The conversion coefficients g_n depend on the active element and the VCO signal.

The non-linear active element is modeled here as depicted in Fig. 4.2. In this way, the necessary odd characteristic with limiting is modeled and the degree of non-linearity can be adjusted by the small-signal gain. The real transfer of the active element is similar but smoother, but the abstraction enables simple calculus and provides more insight. The small-signal transfer function ($\frac{dg}{dx}$) has a constant gain G_M as long as no limiting occurs. Limiting starts at a level K beyond which the small-signal gain is zero. K, i.e. the amplitude must be as high as possible for the lowest phase noise. The influence of the small-signal gain is less transparent; The small-signal gain varies as a square wave with frequency $2f_0$ and amplitude G_M (see Fig. 4.2). The time that the active element behaves as a linear, small-signal amplifier is given by the duty cycle d of the square wave. The duty cycle can be calculated to be:

$$d \cdot \left(1 + \frac{\sin(\pi d)}{\pi d}\right) = \frac{1}{G_M R_p} \qquad (4.14)$$

with $G_M R_p$ the *excess loop gain* . In an ideal, "linear" oscillator, the excess loop gain is one, following the second Barkhausen criterion (Eq. (4.5)). The Fourier coefficient or noise conversion factors for the non-linear oscillator become:

$$g_n = G_M \cdot d \cdot \frac{\sin\left(\frac{n\pi d}{2}\right)}{\frac{n\pi d}{2}} \qquad \text{for } n = \text{even}$$

$$g_n = 0 \qquad\qquad \text{for } n = \text{odd} \qquad (4.15)$$

In a real oscillator the excess loop gain is larger than one to ensure start-up and stable oscillation. This results in more non-linear behavior of the active element, leading to a larger duty cycle of the small-signal gain. For $G_M R_p > 2$, d can be approximated with an error of less than 1.5% by:

$$d \approx \frac{1}{2 G_M R_p} \qquad (4.16)$$

With Eq. (4.15) and Eq. (4.16), a relationship is established between the excess loop gain and the noise conversion factors g_n. Now, the phase noise can be related to the non-linearity introduced by the active element in function of the excess loop gain. The excess loop gain is a parameter set by the oscillator designer, which allows a more optimized oscillator design.

4.3.1.2 Phase Noise Analysis

Using the small-signal model of Fig. 4.2, the influence of the noise in the oscillator on the phase and amplitude stability can be investigated. The noise input to the system E_{ni} is considered white broadband noise, while the noise output of the system E_{no} is considered limited in bandwidth due to the resonator. This assumption is valid in a high-quality oscillator apart from additive white noise that results in the oscillator noise floor.

The general description of the oscillator output signal is $V(t) = (1+a(t)) \cdot \sin(\omega t + \phi(t))$ with $a(t)$ and $\phi(t)$ the low-frequency amplitude and phase noise. Using elementary signal processing theory, the output noise $E_{no}(\omega)$ can be split up in a phase modulating component $E_{no,\phi}(\omega)$ and an amplitude modulating component $E_{no,a}(\omega)$.

$$E_{no,\phi}(\omega_0 + \Delta\omega) = \frac{E_{no}(\omega_0 + \Delta\omega) + E_{no}^*(\omega_0 - \Delta\omega)}{2}$$

$$E_{no,a}(\omega_0 + \Delta\omega) = \frac{E_{no}(\omega_0 + \Delta\omega) - E_{no}^*(\omega_0 - \Delta\omega)}{2} \tag{4.17}$$

In the frequency domain, the resonator transfer function is $H(\omega)$ and the small-signal transfer function of the active element is a convolution with a series Dirac impulses. The transfer function of the total oscillator is then:

$$E_{no}(\omega) = H(\omega) \cdot (E_{ni}(\omega) + E_{no}(\omega)) \cdot \sum_n g_n \delta(\omega - n\omega_0) \tag{4.18}$$

With Eq. (4.17), the small-signal transfer of noise to amplitude and phase noise can be calculated using the following approximations; First, the oscillation frequency is equal to ω_0, which translates to no phase shift in the active element. Second, $H(\omega_0 + \Delta\omega) \approx H^*(\omega_0 - \Delta\omega)$, which is true for an LC-tank. Third, the output noise is band-limited due to the resonator action:

$$H(\omega) \cdot E_{no}(\omega) \cdot \sum_n g_n \delta(\omega - n\omega_0) \approx H(\omega) \cdot (g_0 E_{no}(\omega) + g_2 E_{no}(\omega - 2\omega_0) + g_{-2} E_{no}(\omega + 2\omega_0))$$

$$\tag{4.19}$$

With Eq. (4.6) and Eq. (4.12), the transfer of noise to phase- and amplitude-related sidebands becomes:

$$E_{no,\phi}(\omega_0 + \Delta\omega) = \frac{H_0 \cdot \omega_0}{j2Q\Delta\omega} \cdot \sum_i \frac{g_{-i+1} + g_{-i-1}}{2} E_{ni}(i\omega_0 + \Delta\omega) \tag{4.20}$$

$$E_{no,a}(\omega_0 + \Delta\omega) = \frac{H_0}{1 - (g_0 - g_2) \cdot H_0 + j2Q\Delta\omega} \cdot \sum_i \frac{g_{-i+1} - g_{-i-1}}{2} E_{ni}(i\omega_0 + \Delta\omega)$$

$$\tag{4.21}$$

Assuming that the input noise power spectral density is symmetrical with respect to multiples of the oscillation frequency, the noise power spectral densities can be calculated:

$$S_{no,\phi}(\omega_0 + \Delta\omega) = \left|\frac{H_0 \cdot \omega_0}{j2Q\Delta\omega}\right|^2 \cdot \sum_{i=1}^{\infty} |g_{-i+1} + g_{-i-1}|^2 \frac{1}{2} S_{ni}(i\omega_0 + \Delta\omega)$$

$$S_{no,a}(\omega_0 + \Delta\omega) = \left|\frac{H_0}{1 - (g_0 - g_2) \cdot H_0 + j2Q\Delta\omega}\right|^2 \cdot (g_1^2 \cdot S_{ni}(\Delta\omega) + \qquad (4.22)$$

$$\sum_{i=1}^{\infty} |g_{-i+1} - g_{-i-1}|^2 \frac{1}{2} S_{ni}(i\omega_0 + \Delta\omega))$$

As mentioned in the previous section, all noise around multiples of the oscillation frequency contributes to the phase and amplitude noise; Therefore, the active element should be as weakly non-linear as possible. Note that for a linear, ideal oscillator, only g_0 is non-zero, such that phase and amplitude noise are equal and add up to the well-known linear phase noise formulas.

The above equations are not really transparent. Therefore, phase-conversion factors PM_n and amplitude conversion factors AM_n are defined, which describe the conversion from the noise around $n\omega_0$.

$$PM_n = H_0 \cdot |g_{n-1} + g_{n+1}| \qquad (4.23)$$
$$AM_n = H_0 \cdot |g_{n-1} - g_{n+1}|$$

The power spectral densities reduce to:

$$S_{no,\phi}(\omega_0 + \Delta\omega) = \left[\frac{\omega_0}{2Q\Delta\omega}\right]^2 \cdot \sum_{n=1}^{\infty} PM_n^2 \cdot \frac{1}{2} S_{ni}(n\omega_0 + \Delta\omega) \qquad (4.24)$$

$$S_{no,a}(\omega_0 + \Delta\omega) = \left|\frac{1}{1 - AM_1 \cdot H_0 + j2Q\Delta\omega}\right|^2 \cdot \sum_{n=1}^{\infty} AM_n^2 \cdot \frac{1}{2} S_{ni}(n\omega_0 + \Delta\omega) \qquad (4.25)$$

If the active element discussed in Section 4.3.1.1 is adopted, only the even numbered conversion factors are non-zero due to the odd non-linearity of the limiting function. Further, amplitude balance in the Barkhausen criterion yields $PM_1 = 1$. Since all noise conversion coefficients are known (Eq. (4.15)), the phase-and amplitude conversion factors can be plotted for different values of n versus the excess loop gain (see Fig. 4.3). As can be seen, noise from the third and fifth harmonic contribute to the phase noise for acceptable excess loop gain values while amplitude noise is seriously suppressed.

Since in actual oscillators, the excess loop gain is larger than 2 ($G_M R_p \geq 2$), the phase conversion factors are much larger than the amplitude conversion factors, i.e. $\sum_{n=1}^{\infty} PM_n^2 \gg \sum_{n=1}^{\infty} AM_n^2$, since amplitude noise is suppressed by the limiting of the active element. Therefore the phase noise power spectral density for white input noise $S_{ni}(\omega)$ can be rewritten as:

$$S_{no,\phi}(\omega_0 + \Delta\omega) = \left[\frac{\omega_0}{2Q\Delta\omega}\right]^2 \cdot \sum_{n=1}^{\infty} (PM_n^2 + AM_n^2) \cdot \frac{1}{2} S_{ni}(\omega) \qquad (4.26)$$

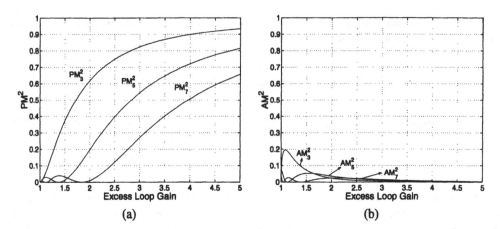

Figure 4.3: *The square of the noise conversion factors for phase (a) and amplitude (b) noise versus the excess loop gain.*

Knowing from Eq. (4.24) and Eq. (4.15) that

$$\sum_{n=1}^{\infty} (PM_n^2 + AM_n^2) = 2H_0 \sum_n |g_n|^2 = 2dG_M^2 H_0^2 \approx G_M H_0 \tag{4.27}$$

then the phase noise power spectral density is given by the simple expression

$$S_{no,\phi}(\omega_0 + \Delta\omega) = \left[\frac{\omega_0}{2Q\Delta\omega}\right]^2 \cdot G_M H_0 \cdot \frac{1}{2} S_{ni}(\omega) \tag{4.28}$$

This means that the phase-related output power is a factor $G_M H_0$, i.e. the excess loop gain, larger than that calculated using a linear method. In this way, a simple relationship is derived between the non-linear transfer characteristic of the active element (i.e. the input-signal or time-dependent behavior of the noise transfer function) and noise conversion from noise around multiples of the oscillation frequency. The equation represents a trade-off; For low loop gains the phase noise power is lowest, but the amplitude noise is less suppressed.

Now it is time to relate the output noise in Eq. (4.28) to the actual circuit parameters and to the phase noise definition of Eq. (2.6). The noise has to be normalized to the carrier power; Since the sinusoidal voltage over the resonator V_A is an easy to measure parameter, the carrier power is defined in rms V^2. To find the wanted units dBc/Hz, the input noise S_{ni} is voltage noise with units V^2/Hz. Looking at Fig. 4.1 (b), it is clear that the current noise sources of R_p (the tank losses) and G_M (the active element) are in parallel and the single side-band voltage noise power spectral density (Nyquist definition) becomes:

$$S_{ni}(\omega) = 4kT R_p^2 \cdot \left(\frac{1}{R_p} + \gamma n G_M\right) \tag{4.29}$$

with k the Boltzmann constant, T temperature and γ the excess noise of the active element ($\gamma = 2/3$ for long-channel MOS devices) and n takes into account bulk transconductance modulation by channel thermal noise ($n \approx 1.5$) [Enz95], meaning that $\gamma n \approx 1$. Since integrated VCOs have a high VCO gain, noise at the tuning input of the VCO also contributes to the output phase noise. For example, if a noisy resistor of 100Ω is present at the tuning input, the resulting phase noise for a K_{vco} of 200 MHz/V is -134 dBc/Hz at 600 kHz. Since the noise spec for DCS-1800 is -116 dBc, the equivalent input noise resistance may not exceed 1.4 kΩ.

Substitution of Eq. (4.29) in Eq. (4.28) yields with some rearrangements:

$$\mathcal{L}\{\Delta\omega\} = 4kT\,R_p \cdot (1 + \gamma n G_M R_p) \cdot \frac{G_M R_p}{2} \cdot \left(\frac{1}{2Q}\right)^2 \left(\frac{\omega_0}{\Delta\omega}\right)^2 \cdot \frac{2}{V_A^2} \qquad (4.30)$$

In an oscillator design, it is sometimes easier to reason with the effective tank resistance R_{eff} in series with the inductor, which is defined by Eq. (4.31).

$$R_{eff} = \frac{R_p}{Q^2} = \frac{(\omega L)^2}{R_p} \qquad (4.31)$$

Eq. (4.30) is easily transformed in terms of this effective resistance:

$$\mathcal{L}\{\Delta\omega\} = kT\,R_{eff} \cdot (1 + \gamma n G_M R_p) \cdot \frac{G_M R_p}{2} \cdot \left(\frac{\omega_0}{\Delta\omega}\right)^2 \cdot \frac{2}{V_A^2} \qquad (4.32)$$

The factor $(1 + \gamma n G_M R_p)$ is the noise factor due to the active element. Due to the excess loop gain $G_M R_p$, the noise factor and the noise multiplication are increased, but amplitude noise is suppressed. The voltage amplitude over the resonator V_A can be expressed as follows; Due to the assumption of a limiting active element transfer, the current through the resonator is a square wave with amplitude I. The voltage amplitude is therefore equal to the current drawn through the resonator by the active element times the resistance of the active element at the fundamental frequency, i.e. $V_A = \frac{4}{\pi} I R_p$. The size of I depends on the implementation of the active element (see Section 4.5.3). In an actual implementation, the amplitude factor for the current is between 1 (sine wave) and $4/\pi$ (square wave). Also the active element transfer of an actual implementation will be smoother and the noise upconversion factor is therefore lower (less coefficients); The phase noise formula above presents a worst-case calculation. If the amplitude formula is substituted in Eq. (4.30), it becomes clear that low phase noise and low power both require a low R_p, which is the main advantage of LC-tank oscillators.

4.3.2 Comparison with Other Published Theories

The most well-known and oldest phase noise theory is the empirical derivation by Leeson [Lees66] of the expected spectrum of an oscillator:

$$S_{\phi,\text{Lees}}(\omega_0 + \Delta\omega) = \left(\frac{\omega_0}{2Q}\right)^2 \cdot \frac{2kT\,F}{P_s} \cdot \Delta\omega^{-2} \qquad (4.33)$$

with F the effective noise figure of the oscillator and $P_s = V_A^2/(2R_p)$ the rms signal power in Watts. By multiplying the signal power by R_p, the signal power in rms V^2 is found. By transforming the double-sideband power spectral density of Eq. (4.33) to single-sideband phase noise, Leeson's formula becomes:

$$\mathcal{L}_{\text{Lees}}\{\Delta\omega\} = 4kT R_p \cdot F \cdot \left(\frac{1}{2Q}\right)^2 \left(\frac{\omega_0}{\Delta\omega}\right)^2 \cdot \frac{2}{V_A^2} \tag{4.34}$$

With $F = 1 + \gamma n G_M R_p$, this formula is equal to Eq. (4.30), apart from the factor $G_M R_p/2$. In fact, Leeson does not make the distinction between phase and amplitude noise, such that his predicted noise is a factor 2 too high, which comes down to an excess loop gain of 2; By including the amplitude noise, Leeson incorporates unknowingly 3 dB noise multiplication due to non-linearities.

Another linear phase noise theory is the one presented by Craninckx [Cran95b, Cran98]. His phase noise formula states:

$$\mathcal{L}_{\text{Cran}}\{\Delta\omega\} = kT R_{eff} \cdot (1 + A) \cdot \left(\frac{\omega_0}{\Delta\omega}\right)^2 \cdot \frac{2}{V_A^2} \tag{4.35}$$

With $A = \gamma n G_M R_p$, the formula reduces to Eq. (4.32), again apart from the factor $G_M R_p/2$, for the same reasons at stated above.

A more general phase noise theory, which takes into account the non-linear behavior of the active element and therefore noise conversion is the theory of Hajimiri [Haji98]. In fact, the phase noise theory presented here is a special case of this theory. In the presented case, the small-transfer function $\frac{dg}{dx}$ is a square wave, while Hajimiri works with the real small-signal transfer function, which he calls the Impulse Sensitivity Function (ISF) $\Gamma(x)$. This ISF is also periodic and can be written as a Fourier series. The noise conversion coefficients g_n reflect the real small-signal transfer function and are non-zero also for odd numbers of n. As a result the predicted phase noise is more accurate. The phase noise calculations do not yield a direct relationship between circuit design parameters and the phase noise and requires a accurate simulations/calculations for each design point. The phase noise formula presented in this book is more straightforwardly applicable by oscillator designers and provides insight on how the active circuit design influences the overall phase noise performance of the oscillator in a single formula by definition of the *excess loop gain*.

A similar theory as the one presented here, though somewhat more elaborate, is the one presented by Samori in [Samo98]. This theory makes distinction between amplitude and phase noise and takes into account noise folding due to the non-linearity of the active element by an additional noise folding factor in the phase noise formula; Larger amplitudes, i.e. more switching, i.e. a higher excess loop gain or non-linear behavior, increase this noise folding factor.

For completeness, the rigorous non-linear phase noise theory of [Demi98] is mentioned. The paper provides a full mathematical phase noise analysis, using Floquet theory to define noise conversion factors. The theory might be mathematically correct, but provides no direct insight in VCO circuit design and is therefore too far off for the common oscillator designer[1].

---

[1] at least for me it is

4.4 Integrated LC-tanks in CMOS

4.4.1 Introduction

From the phase noise formula (Eq. (4.30) and Eq. (4.32)) derived in the previous section, it becomes clear that the quality of the integrated LC-tank is of utmost importance for noise as well as power consumption. Smaller parasitics, i.e. a high R_p or a low R_{eff} means less intrinsic noise, less noise from the active elements and lower power, as shown in the following formulas, taking into account the output voltage amplitude $V_A \propto I R_p$:

$$\mathcal{L}\{\Delta\omega\} \propto \frac{(\omega_0 L)^2}{R_p^3} = \frac{1}{Q^3(\omega_0 L)} = \frac{R_{eff}^3}{(\omega_0 L)^4} = R_{eff}^3 \cdot (\omega_0 C)^4 \tag{4.36}$$

$$G_M = \frac{1}{R_p} = \frac{1}{Q(\omega_0 L)} = \frac{R_{eff}}{(\omega_0 L)^2} = R_{eff} \cdot (\omega_0 C)^2 \tag{4.37}$$

The formulas are derived in different forms to show that one should not jump to conclusions based on the phase noise formulas. For circuit design, it is easiest to reason in terms of Q and R_p. It is clear that the R_p of the LC tank (and therefore the Q) should be as high as possible for noise and power. For design of the LC-tank –the inductor in particular–, the inductance L and its series resistance R_{eff} are of importance. The phase noise formula Eq. (4.32) gives the impression that only minimizing the R_{eff} is enough for low noise. However, from the above equations, it is clear that R_{eff} should be as low as possible (power of 3 in noise), but more importantly, the inductance needs to be as high as possible (power of 4 in noise); In fact, oscillator inductor design comes down to optimizing the Q, with the emphasis on a high inductance. Automatically, the capacitance must be as low as possible for low noise.

One way to realize high-quality inductors with large inductances is gold bonding wires [Cran95a], which are readily available in the IC assembly process. Since gold is highly-conductive, R_{eff} is very low with an the inductance of around 1nH per mm length. The parasitic capacitance is limited to that of the bonding pads. However, due to lack of yield and repeatability (variations of ±6%) the technique is less popular for mass production. In addition, bond wire inductance values are sensitive to shocks and micro-phonic effects.

For monolithic integration, planar spiral inductors laid out in standard metal routing levels of the IC remain the dream of every RF VCO designer. The cost of this solution is lower than that of bonding wire inductors and the die area is reduced with an order of magnitude. On top, the yield and repeatability is almost perfect since processing accuracy ($< 0.1\mu m$) is negligible with respect of the inductor size ($100 \times 100 \ \mu m^2$). Although the integrated planar inductor is the most elegant solution, its quality is limited by several parasitics; The most important are the metal resistance, which increases for higher frequency due to the skin effect and eddy currents, and substrate losses. To decrease the series resistance, multi-level and extra thick metallization, which is a standard feature in most modern deep-submicron processes, are used. To eliminate substrate losses, especially for highly doped substrates, the substrate is selectively etched away underneath the inductor [Rof96a]. However, the extra post-processing steps associated with the etching undermine the main advantage of the planar integrated inductor, i.e. its low cost.

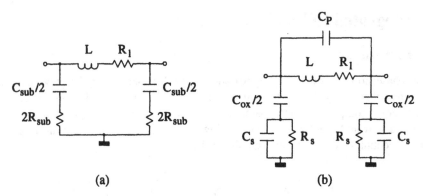

Figure 4.4: *General models for passive integrated inductors: (a) simple and (b) extended.*

Therefore, the next section focuses on the integration and analysis of optimal high-quality planar inductors in standard CMOS technology, without tuning, trimming or extra post-processing steps.

Since the R_{eff} also incorporates the parasitic series resistance of the tank capacitance, the resistance of the varactors which enable tuning of the frequency must be as low as possible for a given capacitance. Different varactor implementations are discussed in Section 4.4.3.

4.4.2   Integrated Planar Inductors in Standard CMOS

4.4.2.1   First Order Planar Inductor Model

Passive integrated planar inductors are mostly realized using a spiral-shaped metal connection, routed in one or more of the standard available metal levels. A general reference for extensive calculations of planar rectangular micro-electronic inductors is given in [Gree74]. The famous Greenhouse article present a method to accurately calculate the inductance value, based on summing or subtracting of the appropriate self- and mutual inductance values of the different segments of the inductor. Calculation of the total inductance value of a full spiral inductor involves the calculation of every mutual inductance between every possible couple of segments. Due to the large number of calculation, the formulas can be embedded in a custom computer program, yielding quite accurate results for the calculated inductance value. Unfortunately, the Greenhouse method does not support segments with a 45° angle and does not provide calculation for the parasitic losses in the inductor.

In [Lee98c], several crude (0th order) calculation are given for integrated spiral inductors. These formulas are only useful for quick hand calculation as a starting point for inductor design prior to verification with a field solver. An example is given for a square spiral inductor:

$$L \approx \mu_0 n^2 r = 4\pi \cdot 10^{-7} n^2 r \qquad (4.38)$$

with n the number of turns and r the radius of the spiral in m. Again, the calculations are only valid for rectangular inductors. To extend the formula to other shapes, it must be multiplied by and area ratio factor, which is $\sqrt{\pi/4} \approx 0.89$ for circular spirals and 0.91 for octagonal spirals. Most of the existing inductance formulas are valid for square spiral inductors, although the Q of circular inductors is higher. This is because most older processes do not support non-Manhattan geometries. Today's processes support $45°$ angles, enabling the integration of octagonal inductors, which better approximate the Q of a circular inductor.

Generally, a complete passive inductor can be modeled by the simple equivalent circuit shown in Fig. 4.4 (a). The inductor L has a parasitic resistance R_l and a capacitance C_{sub} to ground, mainly the oxide capacitance, in series with the substrate resistance R_{sub}. Extensions to the simple model are shown in Fig. 4.4 (b). Shunt capacitance C_P over the inductor mainly takes into account the capacitance of the cross-under metal line to connect the inner turns to the outside world. This capacitance is alleviated by balanced or differential inductor design. C_{ox} is the actual oxide capacitance between the inductor and the substrate. C_s and R_s are the capacitance and the losses of the actual substrate, including loss currents through the oxide and eddy currents induced in the substrate by magnetic fields. This extended model can be fitted onto the measured data of almost every integrated spiral inductor [Nik98].

Note that the definition of the Q of a stand-alone inductor is different from that of a LC-tank. For an inductor, only the energy stored in the magnetic field is of importance, not the electrical energy, which is counterproductive for an inductor. The formula for the Q_L of an inductor, based on Eq. (4.9) for the model in Fig. 4.4 is:

$$Q_L = \frac{\omega L}{R_l} \cdot \left[1 - \left(\frac{\omega}{\omega_0} \right)^2 \right] \tag{4.39}$$

The inductor Q_L is zero at self-resonance.

4.4.2.2 Losses in Integrated Planar Inductors

Metal Losses

For lower operating frequencies, the series resistance of the inductor is the DC resistance of the metal tracks, i.e. the product of the sheet resistance of the metal and the number of squares of the metal tracks. Therefore, multiple metal layers should be used as much in parallel as possible to reduce the DC resistance of the metal. For high-ohmic substrates, all metal layers can be used since the substrate losses are limited (see next paragraph) and the parasitic substrate capacitance is reduced by the substrate resistance. For low-ohmic substrates, depending on the technology, it can be more profitable to omit the bottom metal layer(s), which usually have the highest sheet resistance anyway.

At high frequencies, two effects cause a non-uniform current distribution in the metal tracks, resulting in increased metal losses: the skin effect and eddy currents. The skin effect pushes the current to the outside of the conductor, such that 63% of the current is in a shell with thickness δ, defined as the skin depth. The effective area of the conductor is reduced and hence the resistance increases following Eq. (4.40).

$$R_{l,\delta} \approx \frac{l}{w \cdot \sigma_M \cdot \delta(1 - e^{-t/\delta})} \quad \text{with} \quad \delta = \sqrt{\frac{2}{\omega \mu_0 \sigma_M}} \tag{4.40}$$

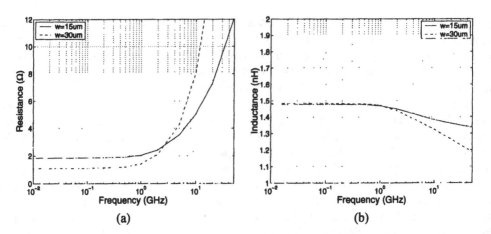

Figure 4.5: *The metal series resistance (a) and inductance (b) versus frequency for two inductors with the same inductance but different conductor widths, without substrate losses.*

with σ_M the conductivity of the metal and l, w and t the length, width and thickness of the metal and μ_0 the magnetic permeability of free space which is very close to that of aluminum and copper. For aluminum used for IC interconnects, $\sigma_{Al} = 3.33 \cdot 10^{-7} 1/(\Omega m)$, such that the skin depth is approximately $\delta_{Al} \approx 1.95 \mu m$ at 2 GHz and $0.87 \mu m$ at 10 GHz. For copper, $\sigma_{Cu} = 5.95 \cdot 10^{-7} 1/(\Omega m)$, such that the skin depth is approximately $\delta_{Cu} \approx 1.46 \mu m$ at 2 GHz and $0.65 \mu m$ at 10 GHz. It is clear that due to the skin effect very wide metal tracks are useless in high-frequency inductors, since only a small part of the conductor is actually used, especially for better conductive material.

A second effect are eddy currents generated in the metal by the magnetic field of the inductor. The situation is depicted in Fig. 4.6 (a) for one side of a inductor with 5 turns carrying a counter-clockwise current $I_{ind}(t)$. This current generated a magnetic field $\vec{B}_{ind}(t)$, which is most intensive in the center of the inductor. The magnetic field is perpendicular to the page and upwards in the center $\odot$. For a fully filled inductor, the magnetic field goes through the inner turns instead of through the center of the inductor. In these turns it induces eddy currents which induce an opposite magnetic field $\vec{B}_{eddy}(t)$ (law of Faraday-Lenz). The counteracting field decreases the inductance at higher frequencies. Moreover, the eddy currents I_{eddy} cause a non-uniform current distribution in the inner conductors; At the outer side of the conductor, the eddy currents cancel the inductor current, i.e. the current is pushed to the inside of the conductor. As a result, at higher frequencies, the resistance of the inner turns increases dramatically. To maintain the inductor quality, "hollow" inductors should be designed, i.e. inductors with no inner turns.

From the discussion, it is clear that these effects are too complex to analyze analytically. Therefore, the metal series resistance is computed with the dedicated planar inductor simulator FastHenry (see Section 4.4.2.3). In Fig. 4.5, the metal resistance and inductance is plotted for two inductors with different widths without taking into account substrate effects. The first inductor has 2 turns, a radius r of $120 \mu m$ and a conductor width w of $15 \mu m$. The second inductor also has 2 turns, but $r = 159 \mu m$ and $w = 30 \mu m$, such that both inductors have approximately the same

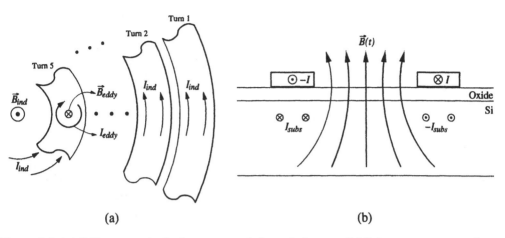

Figure 4.6: *(a) Eddy current in the inner turns of planar inductors. (b) Substrate currents underneath planar inductors.*

inductance. As can be seen, for lower frequencies, the wide inductor has the lowest resistance, but it increases rapidly with higher frequencies and exceeds the metal resistance of the narrow inductor for frequencies higher than 3 GHz. In other words for high frequencies, the conductor width must be limited! For intermediate frequencies (1-2 GHz), the high-frequency effects are less pronounced, making the wider inductor still the best choice. Note that the inductance of the wide inductor also decreases more rapidly for higher frequencies. As a consequence, the Q_L of the narrow inductor keeps increasing for higher frequencies (35 for 20 GHz), while the Q_L of the wide inductor levels out at 10, not taking into account substrate effects.

Substrate Losses

Older CMOS technologies use epi-wafers with a heavily doped, i.e. low-ohmic substrate. Although beneficial for digital design (latch-up, ...), it is a killer for analog design. In highly-conductive substrates, currents induced by the magnetic field of the inductor are free to flow, causing extra resistive losses and a degradation of the inductance. The situation is depicted in Fig. 4.6 (b); The figure gives a typical cross section of a one turn planar inductor on a silicon substrate. The inductor carries a current I at one moment in time (directions denoted by $\odot$, out of page and $\otimes$ into the page), inducing a magnetic field $\vec{B}$. According to Faraday-Lenz, this field induces counteracting currents in the substrate, I_{subs}. The larger the inductor, the larger the magnetic field penetration in the substrate and the more currents, and thus losses are induced. In addition, these substrate currents induce a magnetic field opposite to the magnetic field of the inductor. Therefore, the total inductance of the inductor is reduced.

Fortunately, most of today's CMOS processes (and bipolar and GaAs) use high-ohmic substrates; The magnetic field causes much less substrate currents in these substrates, greatly reducing the effect on the losses in the tank and the reduction of the inductance. Again, an analytic analysis of the substrate losses is cumbersome, if not impossible. Therefore, simulations with a dedicated program such as FastHenry [Fast96] are unavoidable. In Fig. 4.7, the total resistance

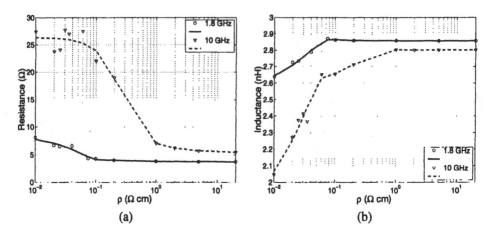

Figure 4.7: *The effect of substrate losses on (a) the inductor resistance and (b) the inductance versus the substrate resistivity at 1.8 GHz and 10 GHz for a inductor with 3 turns, $r = 120\mu$m and $w = 15\mu$m and 300μm leads.*

and inductance of a inductor with 3 turns, $w = 15\mu$m and $r = 120\mu$m and 300μm leads and both metal layer used in parallel is fitted to the simulated data for different ρ's in Ωcm. It must be noted that FastHenry is less accurate for simulation of highly-conductive ground planes, due to the segmented ground plane modeling approach; The accuracy depends on the definition of the ground plane: higher accuracy requires very a long simulation time. For a inductor design on low-ohmic substrates, a final simulation with a finite-element simulator such as Magnet [Mag93] is recommended. However, the plots show clearly that the losses increase by a factor 2 for substrate resistivities lower than 0.1 Ωcm. Simultaneously, the inductance decreases by more than 10%. For higher frequencies, the effect of substrate losses is disastrous for $\rho < 1\Omega$cm; The inductor resistance increases by a factor 6 and the inductance decreases by more than 30%.

Some tricks exist to lower the effect of substrate losses; First, if substrate losses are an issue, it is best to stay away from the substrate, i.e. to use only the top metal layers if possible. Secondly, especially in the case of highly-conductive substrates, the metal layers of the technology can be laid out in series instead of parallel. As a result, the area of the inductor and thus the substrate losses is greatly reduced for the same inductance. The disadvantage is that the DC series resistance of the inductor increases seriously. A third, more elegant solution is the use of patterned ground shields [Yue98]; To cancel energy dissipation in the substrate a short or an open must be realized. The open is using high-ohmic substrates, the short is using shielding. Highly conductive shields in Al (solid or patterned) present low electrical losses, but decrease the inductance by negative mutual coupling. A patterned poly-silicon ground shield is more optimal and increases the Q_L with 33% (at the optimal frequency). The ground shield replaces the lossy, frequency-variant C_{sub} with a larger, but lossless and frequency-invariant capacitance. This reduces the self-resonance frequency of the inductor, but decreases the losses in the substrate, resulting in the higher Q_L. Accurate modeling of the influence of a patterned ground shield is however an outstanding challenge [LeGi01].

In the simple inductor model of Fig. 4.4, the substrate losses are taken into account by R_{sub}. This resistance is necessary to model the influence of the substrate capacitance on losses and the resonance frequency. The influence on the inductor resistance is modeled in R_l. The substrate resistance R_{sub} is hard to calculate and depends on the substrate-type (epi or non-epi, low or high-ohmic) on the geometry and the placing of the substrate contacts. The crude formula to compute the substrate resistance used in this book is [Crol96]:

$$R_{sub} = R_{\square,sub} \cdot \frac{t_{sub}^2}{t_{sub}^2 + A_L} \tag{4.41}$$

with $R_{\square,sub}$ the sheet resistance of the substrate, t_{sub} the substrate thickness and A_L the inductor area. For high values of R_{sub}, the influence of the substrate capacitance and thus the substrate losses are greatly reduced and the self-resonance frequency is enhanced. This is made more clear by transforming the series connection to a parallel RC connection at the wanted frequency. The real substrate capacitance seen by the inductor can be calculated to be a factor $1 + (\omega R_{sub} C_{sub})^2$ lower than C_{sub}. This factor can be quite high (factor 10) for high-ohmic substrates.

Integrated Planar Inductor Design Guidelines

From the above discussion and results obtained from FastHenry, some design guidelines for integrated planar inductors on silicon substrates can be distilled:

- *The width of the metal tracks*: For higher frequencies (above 3 GHz), the width of the metal tracks must be limited; Wider tracks means that the inner turns of the inductor become smaller and contribute less to the inductance, but more to the resistance. To counter this effect, the radius must be increased, which also increases the DC metal resistance. Secondly, the skin effect limits the areal use of the track, such that wide inductors lose their advantage. For intermediate frequencies, wide metal lines can still be more effective than narrower ones (see Fig. 4.5).

- *Spacing between adjacent tracks*: The smaller the spacing, the larger the mutual inductance of the inductor. The typical spacing is 1μm.

- *Hollow inductor design*: At high frequencies, eddy currents blow up the resistance of the inner turns of the inductor, while the inductance contribution is small. They must be omitted, resulting in a *hollow* inductor. At intermediate frequencies and high substrate resistance, filling the inductor has only a minor effect on the performance.

- *The area of the inductor*: For low-ohmic substrates, the area of the inductor must be limited, since the penetration depth of the magnetic field is proportional to the inductor area. The area can be limited by using series connections of the metal layer or by shielding the inductor from the substrate by poly-Si patterned ground shields or using only top metal layers. For high-ohmic substrates ($\rho > 1\Omega$cm), this guideline is only useful for higher frequencies (> 3 GHz).

4.4.2.3 The Simulator-Optimizer

To integrate planar inductors without post-processing steps in plain CMOS technology, the inductors have to be thoroughly analyzed to take into account the above mentioned parasitic effects.

The DC series resistance of the metal can be calculated straightforwardly. However, in many practical applications the extra resistance, added due to substrate spread resistance, skin effect and eddy currents, are in the same order of magnitude of the DC series metal resistance. In addition, the extra losses affect the inductance value. So in principle, a complete electromagnetic simulation is mandatory to accurately model all parasitics. Attempts to mathematically model all inductor parameters are reported in literature, but fail to achieve the necessary accuracy or are not universally applicable [Lee98a, Moh99, Crol96, Nik98].

A typical electromagnetic simulator based on the method-of-moments can perform such simulations, but is too slow for exploring different inductors in a reasonable amount of time (approximately one hour per inductor per frequency) [Lee98a]. Another solution is a full finite-element simulator such as Magnet [Mag93], which is a simulator for magnetic fields in electro-motors. In order to model an inductor accurately, a full three-dimensional simulation is needed, which is impractical, due to speed limitations. A way to circumvent the problem is to exploit the symmetry around the vertical axis of a circular inductor. Only a two-dimensional simulation of a cross-section has to be made to obtain full three-dimensional information [Cran98]. The simulation of one inductor for one frequency still takes about 15 minutes. Due to the long simulation times, performing an optimization is out of the question. In practice, a group of inductors is simulated or even processed and measured, and the best one is empirically determined. These methods are tedious and unable to provide an optimal solution.

To enable accurate and optimal spiral inductor design on silicon substrates, a fast simulator-optimizer program has been developed. The simulator-optimizer employs the fast inductance simulator FastHenry [Fast96]. FastHenry computes the frequency-dependent self and mutual inductances and resistances between conductors. Inductors are specified in FastHenry as a connection of segments with a certain finite conductivity, height and width. The cross section of each segment can be broken up in filaments, each of which carry a uniform current. By increasing the number and placement of the filaments the accuracy of the simulation can be enhanced, at the expense of lower simulation speed and higher memory requirement. The conductive ground planes, which represent the silicon substrate are described in the same way. A typical graphical inductor input to FastHenry is shown in Fig. 4.8 (b). FastHenry is thus optimized for the simulation of planar conductors on planar ground planes. It can take into account the actual octagonal shape, the metal crossings and the leads to which the oscillator circuit is connected. Consequently, accurate results can be obtained in 2-3 minutes, on a HP J2240 machine. For highly conductive substrates ($\rho < 0.1\Omega$cm), FastHenry is less reliable and depends on the definition of the ground plane, which a trade-off between speed and accuracy; Fortunately, most modern CMOS processes have lowly doped substrates with resistance values higher than 5 Ωcm, to reduce substrate noise coupling.

Due to the high simulation speed, FastHenry can be implemented in an optimization loop (see Fig. 4.8 (a)). To achieve a global optimum, a simulated annealing optimization program, more specifically VFSR (Very Fast Simulated Re-Annealing) [Ing89] is chosen . In order to be sure to explore all possible inductor geometries the geometry parameters of the search space should be independent; A three-dimensional parameter space has been defined with parameters : the radius, the number of turns and a normalized radial filling ratio, depending on the width of the turns and the "hollowness".

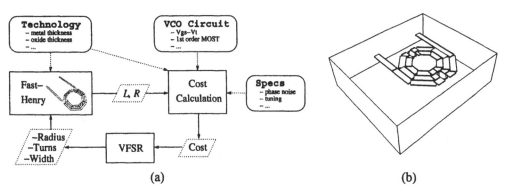

(a)   (b)

Figure 4.8: *(a) The integrated LC-tank VCO simulation and optimization loop and (b) a graphical input for a typical octagonal balanced inductor.*

The inputs for the simulator-optimizer are transistor parameters for a certain $V_{GS} - V_T$ and a certain VCO circuit, technology parameters (metal resistance, oxide thickness, etc.) and the desired phase noise performance, tuning range and operating frequency. The cost function depends on the noise, the tuning range and the power; The noise cost-function is calculated as the normalized difference between the user defined phase noise specification and the actual phase noise performance. The phase noise performance of a certain inductor geometry is derived from Eq. (4.30), using the effective resistance simulated by FastHenry and the given circuit parameters. The tuning cost-function is calculated in the same way as the noise cost-function, but then using the calculated variable-fixed capacitance ratio. The fixed capacitance due to the transistors is calculated from the given transistor parameters. The parasitic capacitance of the inductor is calculated using the technology parameters. The power is estimated using the given transistor parameters and the inductance extracted by FastHenry. The weight of the power spec on the optimization process is adjustable. Due to the weight factor, the simulated annealing algorithm first tries to minimize the cost due to the noise and tuning. When the desired specifications are met (both cost-functions are zero), the power will be minimized. This results in a set of optimal inductor geometry parameters for a certain technology and for a certain VCO circuit.

The rudimentary simulator-optimizer program has been further elaborated and expanded to a fully automated, layout-aware RF LC-oscillator design tool, *CYCLONE* [CYC00] by Carl De Ranter. CYCLONE delivers an accurate and optimal LC-oscillator design, from specification to layout; The tool combines the accuracy of device-level simulation and finite-element analysis with the optimization power of simulated annealing algorithms. Several design experiments show the usability of the tool for a wide range of technologies.

4.4.2.4 The Balanced Octagonal Inductor and Its Model

The inductor geometry employed in all VCO designs in this book, is the balanced octagonal inductor, shown in Fig. 4.9 (b) and (c). Since in most modern technologies, 45° routing is supported, the octagonal shape has been chosen over the more common square shape; Octagonal inductors are closest to circular inductors, which are known to exhibit the highest quality.

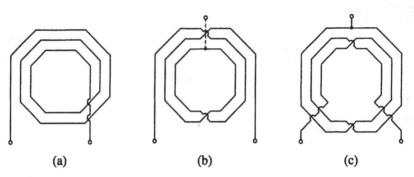

Figure 4.9: *The spiral inductor geometry: (a) conventional octagonal, (b) balanced octagonal and (c) three-terminal balanced octagonal with easy accessible center tap [DeMu00a].*

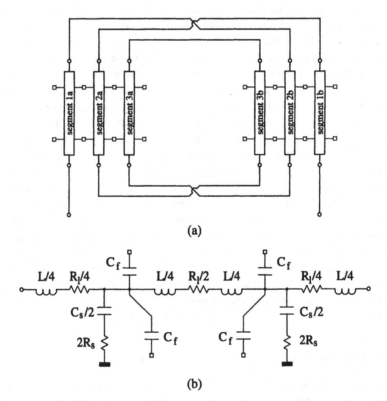

Figure 4.10: *(a) The balanced inductor circuit model and (b) the model of single inductor segment.*

The phase noise of a differential VCO is by construction 3 dB higher than that of a single-ended VCO. Therefore, only differential VCOs are considered, necessitating the use of differential, symmetrical inductors in the LC-tank. The conventional inductor geometry (Fig. 4.9 (a)) is intrinsically asymmetrical, but by placing two of these inductors in series a differential, symmetrical geometry can be realized. However, this topology is very area-expensive and the rather long under-cross connection between both inductors seriously increases the metal losses.

The balanced octagonal inductor has an intrinsic higher inductance for a lower area than the conventional one because of mutual magnetic coupling. It also alleviates the need to study parasitic coupling between two separate inductors. Due to its smaller area, the substrate losses for higher frequencies are reduced in the balanced inductor. Moreover, the self-resonance frequency can be proven to be a factor 2 higher than that of a grounded conventional inductor with the same inductance [Kuhn95]. As a result, the quality factor Q_L of the balanced inductor is significantly better than that of the conventional one (even more than 50% [Nik99]). Fig. 4.9 (b) shows the typical balanced inductor with an optional center tap (dashed line) if more than 2 metal layers are present. For only 2 metal layers, the inductor geometry of Fig. 4.9 (c), presents an easy accessible center tap to realize a three-terminal inductor.

To model the octagonal balanced inductor, a more extended model is used than the one of Fig. 4.4 (a). The inductor is split up in segments that represent one half turn of the inductor as shown in Fig. 4.10 (a). The model of one segment is presented in Fig. 4.10 (b); The inductance is split up in four parts and the resistance in 3 parts. The substrate capacitance C_s, its losses R_s and the fringing capacitances C_f are put in between. The segmented model allows to take into account the distributed nature of the substrate capacitance and inter-turn fringing capacitance; The influence of the both capacitances depends on the voltage over the capacitances, which is not equal to the voltage over the LC-tank. For the balanced inductor in Fig. 4.10 (b), the voltage amplitude decreases from input to the common mode point and increases from there to the other input. The fringing capacitance for a conventional inductor is negligible, since adjacent turns are equipotential; However, for the balanced octagonal inductor of Fig. 4.10 (b), the voltage signal on the outer and inner turns are in phase, but the voltage of the intermediate turn is 180° shifted, since it is situated on the other side of the common mode point. Therefore, a significant voltage difference is present over the adjacent turns, increasing the influence of the fringing capacitances (mainly on the tuning range).

4.4.3   Integrated Varactors in Standard CMOS

To synthesize the specified frequency band under all conditions, the VCO must provide an adequate tuning range. Tuning methods as back-gate tuning [Wang98], special bias techniques [Raz97a] and varying the bias current present only a limited tuning range. Wider ranges are achieved by changing the coupling between LC-tanks [Ping99] or by inductor switching [Kral98], but these methods tend to degrade the phase noise performance. This book is restricted to variable capacitors, a.k.a. varactors, leaving several possibilities for integration in standard CMOS processes: MOS-varactors, accumulation-mode varactors, gated-MOS varactors and P+/N junction diodes.

The MOS-varactor exploits the capacitance change between strong inversion and depletion of a MOS with drain and source connected. In strong inversion, the capacitance and the quality factor are maximal, due to the highly-conductive inversion layer. In depletion, the capacitance is minimal with a high Q. In the transition, the transistor is in weak or moderate inversion, which seriously degrades the quality. The generation of an accumulation layer must be avoided by connecting the bulk to the power supply for a PMOS, since it degrades the tuning for large voltage swings over the varactor. Typical values for the $\xi = C_{max}/C_{min}$ are between 1.56 and 2 [Porr00, Ped99]. The capacitance ratio decreases with CMOS scaling due to the increasing contribution of the fixed drain and source overlap capacitance. Typical values for the Q_{MOS} are around 40 and changes seriously (over 50%) over the tuning. The quality increases quadratically with CMOS scaling as shown in Eq. (4.42) with μ_{mos} the mobility of the minority carriers.

$$Q_{MOS} \propto \frac{\mu_{mos}(V_{GS} - V_T)}{\omega L^2} \tag{4.42}$$

All tuning of the MOS-varactor occurs in a small voltage band, leading to a highly non-linear tuning characteristic [And00, Fig. 9]; Consequently, the VCO gain is high and changes seriously with the output frequency, impeding low-phase-noise PLL design over a large frequency range.

The accumulation-mode MOS-varactor is in fact a PMOS in a N-well, with all P+ contacts replaced by N+ contacts and is operated from depletion to deep accumulation. A PMOS is chosen for the higher mobility of the majority carriers (electrons), leading to a higher Q_{MOS} (see Eq. (4.42)). Again, a dip is present in the Q in the transition between accumulation and depletion, but it scales with CMOS scaling. The generation of an inversion layer must be prevented, since minority carriers are only generated by generation/recombination and therefore scarce, which seriously degrades the quality. Typical ξ values are around 1.7-1.8 [Porr00, Ped99], but the ratio decreases with CMOS scaling in the same way as the standard MOS. The Q of the accumulation-mode MOS is higher than that of the standard MOS, around 40 [Ped99] and 95 [Porr00] and increases quadratically with CMOS scaling (see Eq. (4.42)). All tuning of the accumulation-mode varactor occurs in a somewhat broader voltage band than the MOS, resulting in a somewhat less, but still highly non-linear tuning curve [And00, Fig. 9], with the same unfavorable implications on low-noise PLL design.

An alternative configuration with an enhanced tuning range is the gated-MOS varactor [Wong00]. This structure uses a PMOS with the drain P+ area replaced by a N+ contact. In this way, a combination of a standard MOS, an accumulation-mode MOS and a P+/N junction diode is realized, resulting in an enhanced tuning range. The $\xi = C_{max}/C_{min}$ ratio is higher than 2 for a Q of only 10 to 20. The device has 2 tuning terminals (the drain and/or the gate), which provide a linear tuning range if well combined. However, the rather low Q and the increased complexity to obtain linear tuning make the gated MOS less preferable.

Last but not least, the P+/N junction diode varactor is an option with the N-well as control terminal. Fig. 4.11 shows a balanced configuration. The P+/N diode is operated in reverse and uses capacitance C_d of the depletion area between the P+-diffusion and the N-well as varactor. The losses are concentrated in the lowly doped N-well (resistance R_d). Due to the balanced implementation, a virtual ground plane (dashed) exists between the V_+ and V_- terminals and at V_{ctrl}, such that the parasitic, low-Q well capacitance C_w and its losses are short circuited. The forward biasing of the diode must be avoided, since it degrades the quality of the diode. Unfortunately,

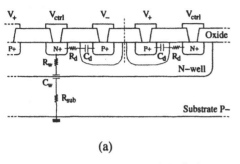

	ξ	Lin	K_{vco}	Q
MOS inversion	+	-	-	$\pm$
MOS accumul.	+	-	-	+
Gated MOS	++	+	+	-
P+/N junction	$\pm^a$	+	+	$+^a$

^{a} depends on layout

(a)   (b)

Figure 4.11: *(a) A cross-section of the balanced P+/N diode varactor implementation and (b) a summary of the performance of different varactor implementations in standard CMOS.*

the capacitance variation is highest in this region and therefore the useful $\xi = C_{max}/C_{min}$ ratio is limited: between 1.32 [Ped99] and 1.82 [Porr00]. The implemented diodes in this book exhibit a ξ ratio, derived from the measured frequency tuning, of 1.93 over a 1.8V voltage range for a 0.25μm CMOS process and 1.56 over a 1.4V voltage range for a 0.65μm (Bi)CMOS process. These values are more than competitive compared to MOS and accumulation-mode MOS varactors and decrease less directly with CMOS scaling. In addition, the quality of the P+/N diodes is high: $Q > 100$ [Porr00, Ped99], but is highly dependent on the layout, which must be optimized to reduce the resistance by e.g. parallelism. The diodes, implemented as in Fig. 4.11 exhibit an estimated Q of over 50. The Q decreases with CMOS scaling if the technology ground rule is scaled accordingly; The minimum spacing between the contacts reduces and hence the parasitic resistance R_d. Most importantly, the Q of a P+/N junction diode is rather constant over the useful tuning range; This results in a more linear tuning characteristic (see [And00, Fig. 9] and Section 4.6.2) and a more constant VCO gain, paving the way to fully integrated, low-phase-noise PLLs in CMOS.

In Fig. 4.11 (b), the performance of the different varactors in standard CMOS are summarized. The P+/N diode has been chosen to implement monolithic, low-phase-noise VCOs in CMOS, for its higher linearity and more constant VCO gain, combined with high Q values and sufficient tuning. However, for high-frequency VCOs (> 5 GHz), the quality of the P+/N diodes drops rapidly with frequency and becomes the limiting factor in terms of noise [DeCo01]. In view of high-frequency design and CMOS technology scaling, accumulation-mode MOS varactors become a more efficient solution, since their Q scales with the square of the gate length, while providing a high capacitance ratio.

4.5 The VCO Circuit Design

4.5.1 General VCO Circuit Design

A general rule for RF oscillator circuit design is: *less is more!* The circuit functionality is best implemented with the least active elements, since each of them introduces extra parasitics

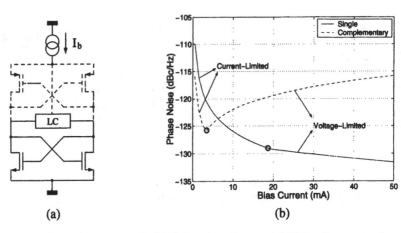

Figure 4.12: *(a) Balanced cross-coupled VCO active element (xMOS-only or complementary). (b) Calculated phase noise at 600 kHz offset versus the bias current for typical LC-tank and transistor parameters and $V_{DD} = 1.8V$ at 2 GHz.*

and losses. The active circuit of an oscillator must provide a negative resistance to compensate the tank losses; One transistor suffices for this task, but for a balanced design minimum 2 transistors are necessary, as shown in Fig. 4.12 (a). The common source structure uses two cross-coupled transistors to provide the required positive feedback. The oscillator output is buffered by common-source buffers to get the signal off chip. Since for oscillator design, MOS transistors provide higher performance than bipolar transistors (see Section 4.5.2), only CMOS VCO circuit implementations are considered in this book. Several implementation flavors exist: NMOS-only, PMOS-only or complementary with NMOS and PMOS cross-coupled transistors (see Section 4.5.3). For biasing, additional transistors are needed which have no influence on the performance in a perfectly balanced design (see Section 4.5.4). The frequency performance is discussed in Section 4.5.2

The VCO design starts with the design of the LC-tank, by finding a geometry that provides a high inductance with the lowest parasitic resistance and capacitance, while maintaining a sufficient tuning range. The parallel tank resistance is extracted and sets the necessary transconductance G_M of the VCO circuit, including some excess loop gain. The excess loop gain is a trade-off between start-up and stable oscillation and noise upconversion (see Section 4.3) and is typically between 2 and 3. In an xMOS-only implementation, the necessary transistor transconductance is $g_{m,x} = 2G_M$. The W/L of the transistors must be limited to have sufficient tuning. Therefore, the overdrive voltage $V_{GS} - V_T$ cannot be too small ($> 0.25V$). However, a higher $V_{GS} - V_T$ means a higher power consumption. The $V_{GS} - V_T$ also influences the amplitude of a xMOS-only circuit when the signal amplitude is voltage-limited, since it determines the DC output voltage of the VCO; The higher the V_{GS}, the higher the oscillation amplitude can be (see Section 4.5.3). In a complementary circuit however, the power supply voltage sets the amplitude in the voltage-limited region. The $V_{GS} - V_T$ also influences the balance of the circuit and thus the (1/f) noise upconversion from the bias current source. In Section 4.5.4, it is shown that an optimal $V_{GS} - V_T$ exist for minimal noise upconversion.

A different transistor sizing approach starts from the bias current I_b; In the current-limited region, the bias current determines the amplitude of the oscillator and not the available voltage (see Section 4.3 and Section 4.5.3). The bias current should be such that the VCO operates at the border of the current-limited and the voltage-limited region, which is the optimal design point for power and phase noise (○ in Fig. 4.12 (b)). In Fig. 4.12 (b), the influence of the current on the phase noise is calculated for an xMOS-only (solid) and a complementary topology (dashed), using Eq. (4.30) and the amplitude formulas of Section 4.5.3. Typical LC-tank ($Q = 10$) and transistor parameters are chosen for a V_{DD} of 1.8V at 2 GHz. For an xMOS-only topology, the phase noise for high currents levels out (the voltage-limited region), since the amplitude is only the square root of the current (through V_{GS}) and the noise upconversion also increases similarly. In a complementary topology, the amplitude is fixed by the power supply and the phase noise increases due to noise upconversion. For low power supply voltages (1.8V in Fig. 4.12 (b)), the lowest phase noise is attainable with a xMOS-only topology at the expense of power.

4.5.2 Bipolar or CMOS?

Silicon bipolar technology is commonly known as a superior compared to CMOS [Sans94] for RF applications; Strangely enough, this is not true for high-frequency, low-phase-noise VCOs.

The only advantage bipolar technology presents for VCO design is its intrinsically lower 1/f noise. As will be explained in Section 4.5.4, any imbalance causes upconversion of white and 1/f noise of the bias current source to phase noise. The lower 1/f noise of the bipolar transistor makes it very suitable for the bias current source. However, the base resistance of the bipolar transistor introduces extra white noise, while the gate resistance of a MOS can be made very small by fingered transistor layout.

The same base resistance r_b limits the operating frequency of bipolar VCOs. The maximum operating frequency of a bipolar VCO circuit can easily be calculated to be $f_{max} = f_T / \sqrt{g_m r_b}$ [Sans94]. In most bipolar technologies, the f_T is maximized by reducing the base width, which increases the base resistance and therefore limits the operating frequency to around one fifth of f_T. This problem is alleviated with SiGe bipolar transistors. In contrast, a CMOS implementation allows very high-frequency operation; There is no f_T limitation, since the gate-source capacitance is tuned out by the inductor. The only limitation is the pole formed by the parasitic resistance and capacitance seen at the VCO output. The resistance includes the drain resistance, which can be rather large in LDD (lightly-doped-drain) structures to counteract hot electrons and the gate resistance, which can be made arbitrarily small by using a large number of fingers in the layout. Therefore, CMOS VCOs exhibit much more high-frequency potential than their bipolar siblings.

Last but not least, the high g_m/I (≈ 40) of a bipolar transistor finishes him off; In bipolar VCO design (NPN-only), when the LC-tank is known, the R_p sets the necessary transistor transconductance (including the excess loop gain) and thus the bias current. Since the necessary bias current is small, the amplitude over the tank is small ($V_A \propto R_p I$) and the phase noise is high. In CMOS VCO design (NMOS-only), the g_m/I is around 6 for a typical $V_{GS} - V_T$ of 0.35V; For the same tank, i.e. the same g_m, the bias current and thus the VCO amplitude is 6.7

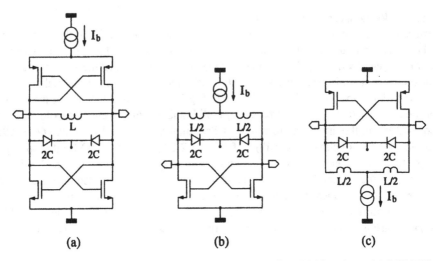

Figure 4.13: *VCO topologies: (a) complementary MOS, (b) NMOS-only and (c) PMOS-only.*

times higher and therefore the phase noise 16 dB lower! If the current in a bipolar VCO is increased for higher signal power, the noise upconversion and intrinsic g_m noise increase linearly with the current and quickly degrade the phase noise (see Eq. (4.30)), while in CMOS these factors increase only with the square root of the current. In short, the power consumption of a bipolar VCO is much lower than that of a CMOS VCO but a tremendous price is paid in phase noise, due to the lower signal amplitude. As long as the VCOs are in the current-limited region, the figure-of-merit (see Section 4.7) of a CMOS VCO is around 8 dB better than that of a bipolar VCO. Even in the voltage-limited regime, the signal amplitude of a bipolar VCO is more limited than that of a CMOS VCO; The output voltage of a bipolar VCO is around 0.7V and therefore the maximum amplitude is limited to around 1.4V. In CMOS, the output is set by the V_{GS} of the MOS transistors and is typically around 1.1V, meaning almost 60% more amplitude (4 dB less noise) and it increases with increasing bias current.

4.5.3 The Power Efficiency of VCO Circuits

4.5.3.1 Hand Calculation MOS Model

In Fig. 4.13, the VCO topologies discussed in this book are shown. These topologies can be modified by removing the current source, which is beneficial for $1/f^3$ noise but is bad for supply pulling. To compute the power efficiency of VCO topologies, a hand-calculation model is used to gain insight in the influence of short-channel effects on VCO design. Actual transistor simulation models (BSIM or MOS level 9) have a high complexity, requiring tens of parameters to model all effects of submicron transistors and therefore provide no design insight. The hand-calculation formula for the current I_{DS} looks like SPICE level 2 and 3 models for a MOS in saturation:

$$I_{DS} = K \frac{W}{L} (V_{GS} - V_T)^2 \cdot \frac{1}{1 - \Lambda V_{DS}} \cdot \frac{1}{1 + \Theta(V_{GS} - V_T)} \qquad (4.43)$$

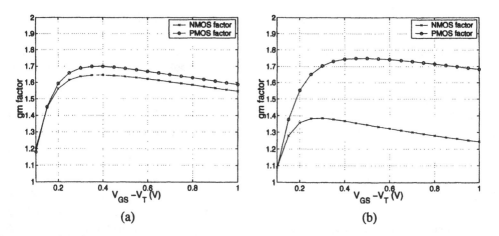

(a)   (b)

Figure 4.14: *The transconductance factor for NMOS and PMOS of (a) a 0.65μm (Bi)CMOS technology and (b) a first-generation 0.25μm CMOS technology.*

type	NMOS					PMOS				
parameter	L_{eff}	K_n	V_{Tn}	Θ_n	Λ_n	L_{eff}	K_p	V_{Tp}	Θ_p	Λ_p
unit	μm	μA/V^2	V	V^{-1}	V^{-1}	μm	μA/V^2	V	V^{-1}	V^{-1}
0.65μm	0.59	60	0.66	1.28	0.07	0.63	27	0.79	1.02	0.07
0.25μm	0.19	218	0.58	4.13	0.11	0.25	43	0.69	0.71	0.07

Table 4.1: *Hand-calculation parameters for two submicron CMOS technologies.*

The formula incorporates the common quadratic relation between current and overdrive voltage, but also Λ for channel length modulation and Θ for mobility degradation. The model is valid for strong inversion and velocity saturation. Θ models both the mobility degradation by the transversal and the longitudinal electric field, similarly as in SPICE level 3. Usually, the longitudinal field effect is much more pronounced than the transversal field effect and it is inversely proportional to the effective gate length. As a result, the mobility degradation factor scales in first order inversely with the effective gate length; This means that for submicron CMOS, the $I_{DS}/(V_{GS} - V_T)$ relation becomes more linear! For the circuits of Fig. 4.13, the V_{DS} is equal to the V_{GS}, but Λ is much smaller than Θ, such that channel length modulation is negligible. The transconductance becomes:

$$g_m = \frac{\partial I_{DS}}{\partial V_{GS}} = \frac{2I_{DS}}{V_{GS} - V_T} \cdot \left[\frac{1 + \frac{\Theta}{2} \cdot (V_{GS} - V_T)}{1 + \Theta \cdot (V_{GS} - V_T)} \right] \qquad (4.44)$$

In Eq. (4.44), the factor between brackets is the *transconductance factor*, which is a measure for the power efficiency of the MOS for a fixed $V_{GS} - V_T$. For long-channel technologies, Θ is small and the transconductance factor is 2, but it decreases with CMOS scaling.

The hand-calculation parameters are extracted using an in-house developed tool, MOSCAL (MOS CALCulator) [MOSC], that performs a least-squares fit to the simulated I_{DS} characteristic of a minimum length MOS in saturation. The K, Θ and V_T are extracted at a V_{DS} of 1.5V. The worst-case error between the extracted hand calculation model and the simulated characteristics is less than 2% for $V_{GS} - V_T$ values between 0.2V and 0.8V. In Table 4.1, the extracted parameter values for the CMOS technologies, employed in this book for VCO implementations, are summarized.

In Fig. 4.14, the transconductance factor is plotted versus the overdrive voltage for NMOS and PMOS of the 0.65μm (Bi)CMOS technology (a) and the 0.25μm CMOS technology (b). From this figure and Table 4.1 it is clear that the mobility degradation, represented by Θ, is much more pronounced in NMOS transistors, while it merely affects PMOS transistors. This is due to the higher mobility of electrons and the lower channel doping of NMOS transistors. The mobility degradation is much higher for smaller gate/channel lengths, i.e. as high as 4.13 V^{-1} for the 0.25μm technology. As a result the transconductance factor is in the rather large 0.65μm CMOS technology similar for NMOS and PMOS, around 1.7. But in the 0.25μm technology, the transconductance factor at useful $V_{GS} - V_T$ values (around 0.35) is much smaller for NMOS transistors, affecting the power efficiency of VCO circuits.

4.5.3.2 Complementary MOS or xMOS-only VCO?

To compare the performance of the different VCO topologies of Fig. 4.13, it is assumed that the bias conditions and the LC-tank are the same for all topologies. Furthermore, the transistors are assumed to fully switch, such that the drain current of the transistors is a square wave.

The VCO design starts again with the R_p of the tank which sets the necessary transconductance of the transistors. In the complementary topology (Fig. 4.13 (a)), the transconductances of the NMOS and PMOS must be equal for balance and the sum of both transconductances is equal to G_M, while for the xMOS-only topology $G_M = g_{m,x}/2$. In the complementary case, the bias current I_b is drawn through the LC-tank twice every period, while in the xMOS-only case the current is drawn through half of the tank. For the complementary VCO, the voltage amplitude over the LC-tank is $V_A = \frac{4}{\pi} R_p I_b$, with $\frac{4}{\pi}$ the Fourier coefficient of the fundamental frequency. The amplitude of the xMOS-only topology is half, $V_A = \frac{2}{\pi} R_p I_b$. This means that the complementary topology has an intrinsic 6dB better phase noise potential for the same current if the amplitude is current-limited. In the voltage-limited region, the amplitude of the complementary VCO is limited by the power supply voltage V_{DD}, while the amplitude of the xMOS-only topology is limited by the V_{GS} of the xMOS transistors.

The performances of both a complementary VCO and a NMOS-only VCO have been calculated in first order; For a given LC-tank ($Q=10$, $R_p = 300\Omega$) at 2 GHz and 1.8 V and the earlier presented 0.25μm CMOS technology, the transconductances have been calculated and the transistors are sized. From then on, the bias current is changed and the influence on the performance is evaluated using Eq. (4.30). The results are shown in Fig. 4.12 (b). In the current-limited region, the complementary VCO performs better since its amplitude is a factor 2 higher. However, when the VCO signal is voltage-limited, the amplitude of the complementary VCO is fixed by the V_{DD}, while the amplitude of the NMOS-only topology still increases with the square root of I_b through V_{GS} (short-channel effects are incorporated); The noise of the complementary VCO increases in the voltage-limited region, since the higher excess loop gain results in higher active element noise and more noise upconversion, while the noise of the NMOS-only still decreases with the current.

From the above calculation, it can be concluded that if the power supply voltage is sufficiently high, the complementary VCO presents the most power efficient solution. As can be seen in Fig. 4.12 (b), it achieves its optimal noise (o) at a much lower current than the NMOS-only VCO. For a V_{DD} of 3V, the optimal phase noise is as low as that of the NMOS-only VCO (around -128 dBc/Hz). However, when the power supply voltage is lower than 2 V, the phase noise levels achievable with the NMOS-only topology are lower, at the expense of a higher current. The necessary power supply voltage for the xMOS-only VCO is approximately 1/4 of that of a complementary one for the same bias current and equal phase noise; This means that the signal power to DC power ratio of both topologies is the same and that for very low power supply voltages, the NMOS-only topology is more power efficient.

4.5.3.3 NMOS-only or PMOS-only VCO?

From the hand-calculation model, it becomes clear that the transconductance factor of a PMOS for smaller technologies is better than that of an NMOS. Therefore, the question should be asked, whether NMOS or PMOS transistors should be used as active elements. In an xMOS-only VCO design, the transconductance $g_{m,x}$ is again determined by the LC-tank and the $V_{GS} - V_T$ is set by the amplitude and 1/f noise upconversion (see Section 4.5.4). From Fig. 4.14 it is clear that for the 0.25μm process, the current needed to implement the necessary transconductance is almost 30% smaller for the PMOS-only VCO (see Fig. 4.13 (c)). Unfortunately, the amplitude is also smaller and therefore the phase noise worse.

In Fig. 4.12 (b), the calculated noise curve of the PMOS-only VCO would be higher than that of the NMOS, but the optimum occurs at a lower bias current. Since the calculations are quite rudimentary, it is advisable to perform simulation to decide which topology to use. For the 0.25μm CMOS process, it has been simulated that the overall performance of the PMOS-only topology is slightly better than that of a NMOS-only topology. Moreover, the useful range of the tuning diodes is extended (see Section 4.6.2). For the 0.65μm, the NMOS-only topology performs better. Moreover, the parasitic capacitance of the NMOS transistors is smaller and the PMOS current source, which has intrinsically lower 1/f noise, can be implemented.

4.5.4 $1/f^3$ Phase Noise Mechanisms and Minimization

Compared to bipolar transistors, CMOS transistors generate more flicker noise, which in oscillators is upconverted to $1/f^3$ shaped phase noise close to the carrier. Consequently, the $1/f^3$ phase noise is higher in CMOS oscillators and can become an issue. Flicker noise upconversion determines the phase noise at small offset frequencies, where it is suppressed when the oscillator is used in a PLL frequency synthesizer. However, when the oscillator is not carefully designed, flicker noise can deteriorate the phase noise at higher offset frequencies, important for communication systems, e.g. 600 kHz by several dBs.

The mechanism of flicker noise upconversion can be explained as follows; When the VCO circuit is unbalanced, the common mode node of the current source oscillates at twice the oscillator frequency, $2\omega_0$, because the current source is pulled every time one of the xMOS transistors switches on. Through channel length modulation, the noise of the tail current source is upconverted to $2\omega_0$. The upconverted noise enters the LC-tank and is mixed with the fundamental frequency, resulting in phase noise sidebands at the oscillator frequency and the 3rd harmonic. Therefore, to minimize the upconversion of flicker and white noise from the tail current source, balance must be preserved, meaning that all even harmonics must be suppressed. Odd harmonics have little importance for flicker noise upconversion because they do not affect the symmetry. Flicker noise from the tail current source is the main contributor to $1/f^3$ phase noise. The contribution of the xMOS g_m-transistors is small due to the switching of the oscillator. Flicker noise is correlated noise and can only exist in systems with memory. When transistors are ideally switched, all memory and consequently the flicker noise is removed [Gier99]. When the switching is not ideal, a small amount of the xMOS transistor flicker noise is upconverted.

The relationship between noise upconversion and symmetry is confirmed by [Post98], which states that flicker noise upconversion is minimal for waveforms with half-wave symmetry. This means that the waveforms satisfy: $f(t) = -f(t \pm T/2)$ with T the waveform period, i.e. the waveforms have no even harmonics. In fact, the relationship between equal up-and down slopes of the waveforms is easily calculated; It is assumed that phase noise is determined at the zero-crossings and the slopes are linear around these points. The phase noise due to amplitude noise v_n is $\theta_n = \frac{2\pi}{T} \cdot v_n \cdot \frac{dt}{dV_A}$ with $\frac{dV_A}{dt}$ the slope of the waveform. In one period, two slopes can be defined, the up-slope $\frac{dV_{A,u}}{dt}$ and the down-slope $\frac{dV_{A,d}}{dt}$. Since the oscillation frequency is fixed, the inverse of both slopes add up to a constant K_f dependent on the frequency:

$$\frac{dt}{dV_{A,u}} + \frac{dt}{dV_{A,d}} = K_f \tag{4.45}$$

Assuming that v_n is uncorrelated and equal in magnitude within a period, the phase noise power in one period T becomes:

$$\theta_n^2 = \left(\frac{2\pi}{T} \cdot v_n\right)^2 \left[\left(\frac{dt}{dV_{A,u}}\right)^2 + \left(\frac{dt}{dV_{A,d}}\right)^2\right] \tag{4.46}$$

The minimum phase noise is found by setting the derivation of θ_n to the slopes zero and results in: $2\frac{dt}{dV_{A,u}} = K_f$. Substituting in Eq. (4.45) yields $\frac{dV_{A,u}}{dt} = \frac{dV_{A,d}}{dt}$. This means that waveform symmetry leads to minimal phase noise.

The phase noise analysis of Section 4.3 takes into account non-linear behavior in the VCO circuit. However, it is assumed that the circuit is balanced, i.e. all odd Fourier coefficients of the signal transfer are zero (see Eq. (4.15)). This means that only noise around odd harmonics is upconverted. The theory does not take into account common-mode noise of the tail current source that can be upconverted due to imbalance. To take into account bias current source noise upconversion, the small-signal transfer $\frac{dg}{dx}$ should be used in its more general form, i.e. with all Fourier coefficients; In fact this comes down to the method presented in [Haji98].

Here, a flicker noise upconversion factor $\Gamma_{1/f}$ is defined, which is the difference between the minimum and maximum of the derivative of the VCO output waveform V_A.

$$\Gamma_{1/f} \propto \max\left(\frac{dV_A}{dt}\right) + \min\left(\frac{dV_A}{dt}\right) \tag{4.47}$$

These derivatives are proportional to the up and down-slopes of the waveform. The difference between the slopes indicates the asymmetry of the waveform and thus its sensitivity to low frequency noise upconversion. The extremes of the derivative occur at the zero-crossings where the oscillator is most sensitive to noise induced phase errors.

To determine the white noise upconversion the general phase noise method of [Haji98] is applied; An oscillator is treated as a linear-time-variant (LTV) system for which an Impulse Sensitivity Function (ISF) $\Gamma(\omega_0 t)$ can be defined. The ISF is a periodic function that describes the sensitivity of an oscillator to phase perturbations due to fluctuations produced by noise in the oscillator. It is a function of the derivative of the oscillators output waveform. In fact, the ISF is similar to the small-signal transfer $\frac{dg}{dx}$ defined in Section 4.3. The ISF is extracted from HSpice simulations. The ISF is defined as follows:

$$\Gamma(\omega_0 t) = \frac{c_0}{2} + \sum_{n=1}^{\infty} c_n \cos(n\omega_0 t) \qquad (4.48)$$

where c_0, which is the average value of the ISF, accounts for the upconversion of low frequency noise (flicker noise) and c_n for the upconversion of noise at multiples of the carrier frequency ω_0. The rms value of the ISF, Γ_{rms}, is proportional to the total upconverted white noise. c_0 can be extracted approximately by calculating the average value of the derivative of the simulated output waveform. Since the presented oscillators are balanced circuits, the output waveforms are quite symmetrical and consequently, the DC value of the derivative is very small. As a result, the accuracy of c_0, when extracted from HSpice simulations, is highly sensitive to the accuracy (number of points, number of periods) of the simulation and choice of the start and end points of the simulated waveform, which must be exactly the same to obtain good results. $\Gamma_{1/f}$ is less sensitive to simulation inaccuracy and is fast and easy to extract from simulations. Therefore, it is applied as the measure for the flicker noise upconversion instead of c_0.

Based on the ratio of flicker noise and white noise upconversion factor, a factor γ_{1/f^3} proportional to the $1/f^3$ phase noise corner frequency can be calculated, $\gamma_{1/f^3} \propto \Gamma_{1/f} / \Gamma_{rms}$. The developed method provides a fast way to examine different design trade-offs to minimize flicker noise upconversion based on HSpice simulations. It is used in Section 4.6.1.2 to determine the bias point for minimal $1/f^3$ phase noise of VCOs and to evaluate further flicker noise upconversion minimization by balancing circuit techniques. It must be noted that the $1/f^3$ noise corner depends on the ratio of white and flicker phase noise and is not equal to the device $1/f$ noise corner!

Another $1/f^3$ phase noise mechanism, which in fact does not originate from $1/f$ noise is parametric noise [Eur93]. Since the large sinusoidal VCO signal changes the varactor capacitance within one oscillation period, it also affects the instantaneous oscillation frequency. Due to the signal the capacitance changes $C_0(t) = C_0 + C(t)$ and thus the frequency changes $f_0(t) = f_0 + \Delta f(t)$ in which $\Delta f(t)$ is noise on the frequency. This noise usually has a flicker noise-like power spectral density $S_{\Delta f}(f) \propto 1/f$, since an increase in capacitance means a decrease in frequency. As a result, the parametric noise merges with the flicker noise of the bias current source. The parametric noise and its relative contribution to $1/f^3$ noise is hard to predict and is only mentioned to show its existence.

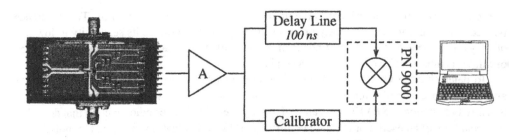

Figure 4.15: *The VCO phase noise measurement setup with the delay-line method.*

4.6   Implementations

All phase noise measurements presented in this section are performed with a dedicated phase noise measurement setup; First, the VCO ICs are glued and wire bonded onto a thick film ceramic substrate. The two VCO RF outputs are buffered by on-chip common-source buffers and connected to the outside world with SMA connectors. SMD capacitors provide extra off-chip decoupling of the power supply. Finally, the substrate is mounted in a Copper-Beryllium box, that serves as the common ground and provides (limited) shielding of the VCO IC.

A single-ended VCO output is directed to the phase noise measurement system, the PN-9000 of Europtest, shown in Fig. 4.15. The delay-line method is used, since a stand-alone VCO is not a stable frequency source and can therefore not be locked onto an external frequency reference. Therefore, the VCO signal itself is used as the reference; The DUT signal is first amplified to compensate for the losses in the delay line and then split up by a power splitter. One signal path passes through the delay line, the other through the calibrator; The calibrator is in fact a phase shifter to ensure that the input signals to the mixer are 90° for quadrature lock. The two signals downconvert each other to baseband and the actual phase noise spectrum is computed. The mixer is insensitive to amplitude noise, such that the measurement reflects the actual phase noise. The delay line of 100ns decorrelates the noise of the two signal paths for measurement accuracy. The delay is short enough to ensure that the frequency variations of the VCO show up simultaneously at both mixer inputs, allowing the system to maintain quadrature lock. The system performs accurate measurements for signals between 400 MHz and 1.8 GHz. The frequency range is extended using low-noise, microwave mixers. A pole in the delay line impedes accurate phase noise measurements for offset frequencies above 5 MHz.

4.6.1   A 2 GHz Low-Phase-Noise LC-VCO Set with Flicker Noise Minimization in 0.65μm (Bi)CMOS

4.6.1.1   Inductor Design

Two separate low-phase-noise VCOs are designed to demonstrate the feasibility of very low-phase-noise performance in standard CMOS technology and to present a technique to lower the

Substrate thickness	t_{sub}	280 μm
Substrate resistivity	ρ_{sub}	8 Ωcm
Field oxide thickness	t_{ox}	1.5 μm
Metal1 sheet resistance	$R_{\square,M1}$	95 $m\Omega/\square$
Metal2 sheet resistance	$R_{\square,M2}$	40 $m\Omega/\square$
Metal3 sheet resistance	$R_{\square,M3}$	40 $m\Omega/\square$

Table 4.2: *The most important physical technology parameters of the 0.65μm (Bi)CMOS process.*

# turns	3
Width (μm)	55
Radius (μm)	191
L (nH)	1.64
R_l (Ω)	1.54
C_{sub}(pF)	3.78
R_{sub}(Ω)	115
Q_L	11

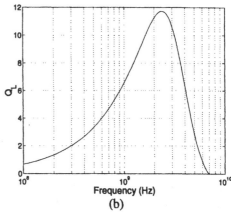

(a)   (b)

Figure 4.16: *(a) The optimized inductor parameters in the 0.65μm CMOS technology. (b) The simulated Q_L versus frequency .*

noise upconversion of the bias current source, which is identified as the main source of $1/f^3$ phase noise (Section 4.5.4). In both designs, the simulator-optimizer of Section 4.4.2.3 has been deployed to explore the boundaries of the technology. The optimization goal is an inductor with a very low series resistance and a reasonable inductance value. Based on the simulations, a three-terminal balanced octagonal inductor is implemented in a 0.65μm BiCMOS process. The most critical technology parameters are listed in Table 4.2. The three available metal layers are laid out in parallel to minimize the DC series resistance. The optimized inductor parameters are listed in Fig. 4.16 (a), with the electrical parameters referred to the simple model of Fig. 4.4 (a). The extra inductance and resistance of the leads has been taken into account.

No non-standard techniques, such as etching, extra thick top metal layers, thick field oxide or special substrate doping, are employed to improve the quality of the integrated inductor. The inductors series resistance is very small due to the optimization process. For low power, the inductor with the highest inductance is chosen by the simulator-optimizer, while satisfying the user-defined phase noise constraint. While one of the inductor design guidelines (Section 4.4.2.2) was not to use wide metal tracks, the optimization resulted in a width of 55μm. Due to the wide tracks the DC series resistance is very low (0.66Ω); Although the parasitic resistance increases rapidly with frequency as shown in Fig. 4.5 (a), the wide metal tracks still present a lower para-

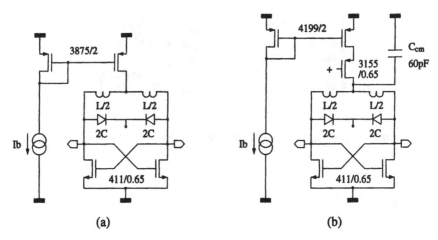

Figure 4.17: *Circuit schematics for both VCOs: (a) The conventional VCO (VCO1) and (b) the VCO with flicker noise minimization (VCO2).*

sitic resistance at 2 GHz. Another discrepancy with the design guidelines is that the inductor is not hollow; The optimisation process did not include varying metal track widths. To make the inductor more hollow to limit eddy current generation in the inner turns, the inner turns can be made more narrow towards the center of the inductor. This however seriously increases the DC series resistance and is therefore not implemented. Due to the rather high substrate resistance, the area is not limited by substrate losses and by parasitic capacitance. At the center frequency ω_0 , the RC-series connection can be transformed to a RC-parallel connection. The actual capacitance seen by the inductor at one side can be calculated to be only 61 fF. Therefore, in the case of high resistive substrates, the tuning range is not limited by the inductors parasitic capacitance. The Q_L is simulated using the model discussed in Section 4.4.2.4 and is plotted in Fig. 4.16 (b). The maximum Q_L is at 2.4 GHz and is 12, while at 2 GHz, the quality is around 11. The self-resonance frequency is 7 GHz.

4.6.1.2   VCO Design with Flicker Noise Minimizaion

The two integrated VCO circuits are shown in Fig. 4.17. Both VCOs are designed for 2 GHz operation. For low power supply voltage, the NMOS-only topology is chosen over the complementary differential topology to enable the oscillators to operate in the current limited region. At a power supply voltage of 1.8V, the maximum differential amplitude that can be obtained with complementary topologies is approximately 1.5V. Balanced NMOS-only topologies can obtain amplitudes, that exceed the power supply (up to 2.3 V (see Fig. 4.18 (b)). The inductor, discussed in the previous section, is implemented in both VCOs. The tuning of the VCO is done by variable capacitances, consisting of two P+/N-well diode junctions of around 6.5pF each for a control voltage of 1.8V. The fixed capacitance at one side of the VCOs is approximately 1.1 pF of the cross-coupled NMOS transistors, 100 fF of the common-source buffers and 61 fF of the

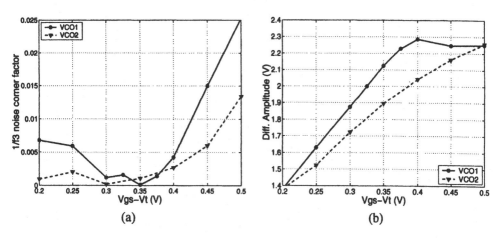

Figure 4.18: *(a) The simulated $1/f^3$ phase noise corner factor γ_{1/f^3} and (b) the simulated differential amplitude versus the NMOS cross-coupled pair gate overdrive voltage.*

inductor.

The VCO design is again started by defining the necessary transconductance of the VCOs. The total effective resistance of the VCO, taking into account diode and transistor parasitics is simulated to be 1.88Ω, 80% of which is attributed to the losses in the inductor. This means that the parallel tank resistance R_p is 226Ω at the highest frequency. Therefore, the necessary G_M is around 5mS and the excess loop gain is designed to be nearly 3 at 2 GHz to ensure stable oscillation and fast startup over the full frequency range of the oscillator.

The most important degree of freedom in CMOS VCO circuit design is the gate overdrive voltage, the $V_{GS} - V_T$, of the NMOS transistors. The choice of the $V_{GS} - V_T$ determines the output amplitude, the power consumption, the added parasitic capacitance of the NMOS transistors and the flicker noise upconversion. Fig. 4.18 shows the $1/f^3$ corner factor γ_{1/f^3}, determined with the procedure explained in Section 4.5.4 and the VCO output amplitude versus the NMOS overdrive voltage. The values where extracted from Hspice transient simulations, keeping the g_m of the NMOS transistors constant. A minimum exists for the $1/f^3$ phase noise corner factor at 0.35 V. However, at that bias point, the differential amplitude (see Fig. 4.18 (b)) is only moderate. As a trade-off, a $V_{GS} - V_T$ of 0.375 V was chosen. This sets the simulated bias current I_b to around 15mA.

As stated in Section 4.5.4, the main source of upconverted flicker noise is the bias current source. A PMOS transistor was chosen for the bias current source, for its inherently lower flicker noise (approximately 10 dB), compared to NMOS transistors. The area of the PMOS transistor is increased, by designing the transistor length larger than the minimum length, to further minimize the flicker noise contribution. Its $V_{GS} - V_T$ is 0.5V to increase the accuracy of the current mirror. A mirroring factor of 1:1 has been chosen to make sure that the 1/f noise of the mirror transistor does not nullify all attempts to minimize the flicker noise.

In VCO2, additional circuit techniques are applied to further minimize the flicker noise upconversion. A capacitance and a cascode transistor are added to the common mode node of the

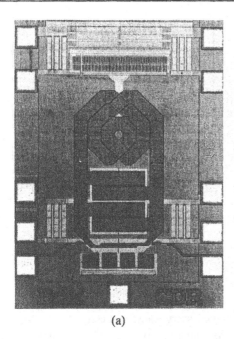
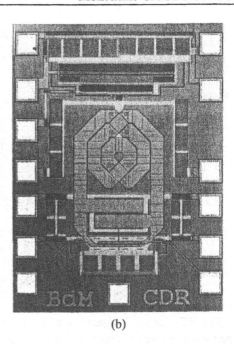

(a)   (b)

Figure 4.19: *IC microphotographs for both VCOs: (a) The conventional VCO (VCO1) and (b) the VCO with flicker noise minimization (VCO2).*

bias current source to suppress all common mode node variations. As a result, noise upconversion due to channel length modulation by higher (and especially the second) order harmonics is suppressed. The capacitor C_{cm} of 60 pF provides a low impedance path to AC ground for higher harmonics. The cascode transistor increases the output resistance with an extra $g_m \cdot r_o$, further reducing the common mode node variations. The resulting $1/f^3$ phase noise upconversion factor and differential amplitude versus overdrive voltage are shown in Fig. 4.18. At lower and higher overdrive voltages, the flicker noise upconversion is substantially smaller due to the balancing circuit techniques. The differential amplitude is however lower; The effect of the balancing circuit techniques is a more linear VCO, resulting in an amplitude factor closer to one (1.09), i.e. a more sinusoidal signal for VCO2 when compared to VCO1 with a factor closer to $\frac{4}{\pi}$ (1.21). At high amplitudes (high overdrive voltages), both VCOs operate in the voltage-limited region. Due to its overall lower output amplitude, VCO2 has a worse phase noise performance than VCO1 at higher offset frequencies. The phase noise values of both VCOs can be calculated using Eq. (4.30), resulting at 600 kHz offset and 2 GHz in -125.9 dBc/Hz for VCO1 and -123.1 dBc/Hz for VCO2.

4.6.1.3   Experimental Results

The maximum oscillation frequency of the VCOs is somewhat higher than 2 GHz. The tuning range is 11%, 1.79 to 2 GHz for a tuning voltage range of 1.1-1.8 V. Microphotographs of both

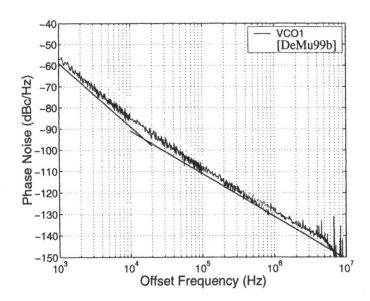

Figure 4.20: *Phase noise measurement of VCO1 compared to the phase noise of the 1.33GHz VCO implemented in the same 0.65μm (Bi)CMOS technology [DeMu99b].*

VCO circuits are presented in Fig. 4.19. Fig. 4.20 shows the measured phase noise spectrum of VCO1 (black curve). The black lines illustrate the -20dB/dec and -30dB/dec phase noise slopes. The phase noise is measured to be -125.1 dBc/Hz at 600 kHz and -138 dBc/Hz at 3 MHz. The measured phase noise exceeds the most stringent specifications for DCS-1800 systems (Section 2.6). The excellent phase noise performance was achieved at a power consumption of 19 mA from a 1.8 V power supply. To show the dependence of the phase noise on the NMOS transistor overdrive voltage, both VCOs were measured at different bias currents (see Fig. 4.21) and the result are re-calculated to overdrive voltages. The phase noise gets better for higher overdrive voltage due to the higher amplitude, as long as the oscillators operate in the current-limited region. At $V_{GS} - V_T$ values higher than 0.4 V, the VCOs enter the voltage-limited region. At 0.475 V, the measured phase noise of VCO1 is as low as -127.5 dBc/Hz (for 60mW). The calculated and measured phase noise are in close correspondence (within 1dB).

The phase noise difference between the two VCOs is larger than predicted by simulations and VCO2 enters the voltage-limited region earlier than expected. Both effects can be explained by the fact that changing the bias current, changes the g_m of the NMOS transistors. When the bias current is increased, the resulting amplitudes are higher than simulated. Due to the balancing circuit techniques, this effect is less pronounced in VCO2. VCO2 will enter the voltage-limited region sooner than expected because of the higher amplitude and the large saturation voltage, needed to keep the cascode current source in the saturation region. At overdrive levels smaller than 0.3 V, the VCOs fail to oscillate because of the too low transconductance. Fig. 4.21 (b) shows the measured $1/f^3$ phase noise corner versus the overdrive voltage. VCO1 has a phase noise corner of 15 kHz, measured at the same bias point as in the previous section. The flicker

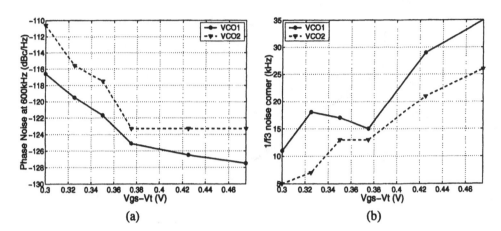

Figure 4.21: *(a) The measured phase noise at 600 kHz offset and (b) the measured $1/f^3$ phase noise corner frequency versus the NMOS gate overdrive voltage.*

noise upconversion is smaller for VCO2, especially at the highest and lowest overdrive voltages as shown in the simulations. The $1/f^3$ phase noise corner differences are rather small, because both VCOs are already optimized, so that additional circuit techniques can only have a small impact. To show that the technology is not responsible for the good flicker noise upconversion results (gray curve), the phase noise performance of the VCO in [DeMu99b] is presented in Fig. 4.20, together with the phase noise of VCO1. The VCO is implemented as a test case for the simulator-optimizer and operates at 1.33 GHz. Its phase noise is as low as -127.5 dBc/Hz at 600kHz for a power consumption of 20mW. For comparison, the phase noise of the oscillator is scaled to 2 GHz. The VCO is implemented in the same (Bi)CMOS technology, but since its bias point is not optimized, the measured $1/f^3$ phase noise corner is around 200 kHz.

At a small frequency band around 2.02 GHz, the measured phase noise spectrum of VCO2 showed no $1/f^3$ phase noise at all! Moreover, the measured phase noise at 600 kHz offset was lower than -132.5 dBc/Hz, due to bias current filtering. The measurement is shown in Fig. 4.22 (b). The capacitor C_{cm} acts together with the bondwire inductances as a filter for the bias current source. At resonance, the impedance is zero and the common mode node is a perfect virtual ground, such that noise from the current source is removed. A more elaborate discussion is given in Section 4.6.1.4.

In Table 4.3, the measured phase noise and the measured $1/f^3$ phase noise corner frequency are compared with state-of-the-art fully integrated (Bi)CMOS VCOs. For comparison, the phase noise is recalculated to 1.8 GHz at 600 kHz offset. As can be seen, the phase noise of this VCO is among the best published; The phase noise at the common mode node filter frequency exceeds that of all published VCOs and no $1/f^3$ phase noise is present at all.

The VCO presented in [Heg01], is in fact the dual principle of what happens with the common node capacitance-bonding wires filter. Instead an inductor is placed between the common mode node and the bias current source. The inductor-capacitor combination acts as a filter at $2\omega_0$, presenting a very high impedance at the common mode node and thereby suppressing all

	ω_0 GHz	$\mathcal{L}\{600\text{kHz}\}$ dBc/Hz	$1/f^3$ corner kHz	Power mW
[DeMu99b]	1.33	-124.8	200	20
[Cran97a]	1.8	-122.5	80	11
[Kin98a]	4.7	-113.9	40	11
[Raz97a]	1.8	-101.6	200	7.6
[DeMu00a]	1.8	-127.5	500	32.4
[Heg01]	1	-130 [a]	100 [a]	10
VCO1/2	2	-126	15/12	34.2
VCO2	2.02	-132.5 [a]	$\ll 10$ [a]	34.2

[a] only valid in a small frequency band

Table 4.3: *A comparison of the phase noise and the $1/f^3$ noise corner of different (Bi)CMOS VCOs with integrated inductors. At the common mode node filter frequency the noise of VCO2 is exceeds that of all published fully integrated VCOs and no flicker phase noise can be distinguished.*

second harmonics at this node. In fact, the current source is made "ideal" at $2\omega_0$ for an "ideal" AC ground. Moreover, the bias current source noise is directed to ground by the capacitor. As a result, flicker noise as well as white noise of the bias current source is suppressed leading to a somewhat lower $1/f^3$ phase noise corner and lower white phase noise noise. Unfortunately, while this technique performs very well at a the frequency where the current source filter is in resonance, the $1/f^3$ phase noise corner frequency quickly degrades to 70kHz, when the VCO is tuned over its frequency range [Hosh01].

An in-house developed technique to minimize noise upconversion is presented in [Borr98]; Here, it is suggested to implement an active feedback loop, that measures the low-frequency signal at the common mode node and feeds it back to the current source, such that the common mode node becomes an "ideal" virtual ground. Consequently, noise from the bias current source is impeded to enter the tank, resulting in better phase noise performance. Last but not least, the most simple trick to get rid of the noise of the bias current source is by simply omitting it [Lev02]. The technique results in a higher power consumption and a worse supply pulling (sensitivity of the frequency to supply voltage variations), but seriously lowers the $1/f^3$ phase noise corner frequency.

4.6.1.4 Extremely Low-Phase-Noise Measurement at 2.02 GHz: -132.5 dBc/Hz at 600 kHz and No Flicker Noise Upconversion !!

In a small frequency band around 2.02 GHz, the measured phase noise of VCO2 is extremely low and no upconverted flicker noise can be distinguished. The measurement is shown in Fig. 4.22 (b). The solid line shows the -20 dB roll-off of white phase noise, showing no -30 dB roll-off, i.e. no $1/f^3$ phase noise at all! The small spurs originate from the computer screen of the measurement setup.

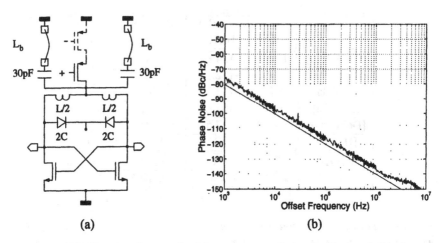

Figure 4.22: *(a) VCO2 as implemented with common mode node filter and (b) the measured phase noise at 2.02 GHz. No upconverted flicker noise can be distinguished and the phase noise at 600 kHz is* −132.5 dBc/Hz *for 1.8V and 18 mA*

The excellent result can be explained as follows; In VCO2, a capacitance C_{cm} was placed at the common mode node of the VCO. For symmetry the capacitance is distributed over both sides and connected to the power supply by bonding wires as shown in Fig. 4.22 (a). The series connection of the capacitance and the bond wire acts as a filter on the common mode node. At the resonance frequency of this filter, the common mode node is a perfect AC ground and is in fact connected to the power supply, i.e. a very low impedance. In other words, the bias current source is short circuited at the resonance frequency in AC and all bias current source noise is shorted to AC ground. As a result, the phase noise spectrum is clean, which confirms that almost all 1/f noise comes from the bias current source (Section 4.5.4).

The frequency band in which the filtering occurs is small due to the high Q of the bonding wire inductors. For further implementations, integrated low Q inductors can be used to broaden the useful frequency band, but this degrade the effectiveness of the noise suppression. A tuning mechanism that follows the VCO tuning can be implemented.

In fact, this filtering method is the dual principle of the filtering proposed in [Heg01]; Instead of a very *low* impedance, the filter at the bias current source in [Heg01] presents a very *high* impedance at $2\omega_0$. The current source is made "ideal" at $2\omega_0$ for an "ideal" AC ground and the current source noise is directed to ground by the capacitor. The filter inductor is implemented on-chip, providing a lower Q and consequently a higher bandwidth but lower performance. As shown in Table 4.3, the white phase noise is low but the $1/f^3$ phase noise corner frequency is still around 100 kHz. Note that, although not mentioned in the paper, this technique only works in a small frequency band but deteriorates fast for other output frequencies.

Substrate thickness	t_{sub}	625 μm
Substrate resistivity	ρ_{sub}	5 Ωcm
Field oxide thickness	t_{ox}	0.7 μm
Metal1 sheet resistance	$R_{\square,M1}$	50 $m\Omega/\square$
Metal2 sheet resistance	$R_{\square,M2}$	35 $m\Omega/\square$
Metal1 thickness	t_{M1}	0.6μm
Metal2 thickness	t_{M2}	1μm

Table 4.4: *The most important physical technology parameters of the 0.25μm CMOS process.*

4.6.2   A 1.8 GHz Highly-Tunable Low-Phase-Noise VCO in 0.25μm CMOS

4.6.2.1   Inductor Design

The simulator-optimizer of Section 4.4.2.3 has been deployed again to integrate a three-terminal balanced octagonal inductor in a 0.25μm standard CMOS technology. Since only 2 metal layers are available in this technology the inside-out balanced inductor topology of Fig. 4.9 (c) has been implemented. The optimization goal is an inductor with a low series resistance for a high inductance value. The most critical physical technology parameters are listed in Table 4.4. The substrate resistivity is 5 Ωcm, which is rather low compared to other commercial deep-submicron CMOS technologies. The technology provides only 2 normal sized metal layers (M1 is 0.6mm thick and M2 is 1mm). The available metal layers are laid out in parallel to minimize the DC series resistance. The inductor parameters are listed in Fig. 4.23 (a), with the electrical parameters referred to the simple model of Fig. 4.4 (a). The extra inductance and resistance of the 300μm leads to connect the inductor to the VCO circuit has been taken into account.

The resulting inductor has an inductance of 2.857 nH and resistance of only 3.7 Ω, using the 2 normal sized metal layers in parallel with abundant via connections. No non-standard techniques, such as etching, extra thick top metal layers, thick field oxide or special substrate doping, are employed to improve the quality of the inductor. The simulated Q_L is plotted in Fig. 4.23 (b); At 1.8 GHz the inductor Q_L is around 8, with a maximum of 11 at 3.5 GHz. The self resonance frequency is 8 GHz. The resulting inductor is clearly distinguishable on the IC microphotograph (Fig. 4.24 (b)); The optimization process has resulted in a inductor, satisfying the earlier defined inductor design guidelines: it is hollow, with metal tracks of only 15μm wide and a moderate area of 120μm. The substrate resistivity is high enough to limit the influence of substrate losses, posing no real limit on the inductor area as shown in Fig. 4.7 (a). The actual parasitic capacitance seen at one side of the inductor can be calculated to be only 270fF.

4.6.2.2   VCO Design

The VCO design is aimed for implementation in a monolithic synthesizer for DCS-1800. The VCO is critical for the tuning and phase noise performance of the synthesizer and should exceed the DCS-1800 requirements. In Table 4.5, the part of Table 2.2 that is determined by the VCO is repeated. Although the capacitance seen by the LC-tank is decreased by the series substrate resistance, when combined with the transistor-added capacitance, it severely decreases the addi-

# turns	3
Width (μm)	15
Radius (μm)	120
L (nH)	2.857
R_l (Ω)	3.7
C_{sub}(pF)	1.83
R_{sub}(Ω)	75
Q_L (1.8GHz)	8

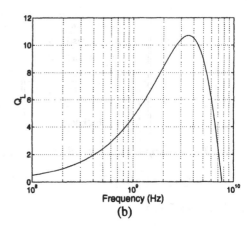

(a)   (b)

Figure 4.23: *(a) The inductor parameters in the 0.25μm CMOS technology. (b) The simulated Q_L versus frequency .*

Phase Noise		
at 600 kHz offset	-116 dBc /Hz	Section 2.6.3.2
at 3 MHz offset	-133 dBc /Hz	
Center Frequency	1.8 GHz	Section 2.6.2
Tuning range	170 MHz	Section 2.6.2
Power Consumption	*ALAP*	Section 2.6.1
Area	*ALAP*	Section 2.6.1
Supply Voltage	1.8V-2V	Section 2.6.1

Table 4.5: *Summary of the VCO specifications in a DCS-1800 communication system.*

tional variable capacitance that can be integrated and therefore the tuning range. To obtain a high tuning range, when working with a low-voltage power supply, a high $\Delta C/\Delta V$ function must be realized. As discussed in Section 4.4.3, MOS-variable capacitors can implement such function, but their tuning is highly non-linear and the resulting VCO gain is large and not constant; This increases the sensitivity of the VCO to voltage noise induced phase noise and obstructs robust PLL design over an extended frequency range. P+/N-diode varactors are better for noise but limit the tuning range. In brief, a wide tuning range and low-phase-noise values are contradictory demands, especially at a low power supply voltage.

One way to achieve a wide tuning range at moderate VCO gain, is to exploit the total available voltage range, i.e. from 0 V to the power supply voltage. Common differential VCO topologies, use NMOS cross-coupled pairs, either combined with PMOS pairs (see Fig. 4.13 (a) and (b)). The DC output voltage of this type of oscillators is typically quite high, due to the required $V_{GS} - V_T$. In order to find an optimal trade-off between a high amplitude and a low power consumption, overdrive levels of 0.3 .. 0.5 V are chosen, resulting in 1.0 .. 1.2 V DC output. When tuning with P+/N diodes with cathodes connected to the control voltage, the diodes start to be forwardly biased when the tuning voltage reaches around 0.5 V. Forwardly biased diodes

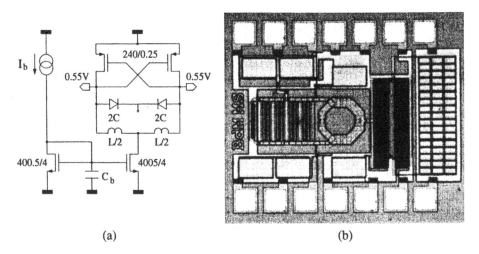

(a)   (b)

Figure 4.24: *(a) Circuit schematics and (b) IC microphotograph of the VCO in 0.25μm CMOS.*

severely degrade the oscillators phase noise, due to extra noise and amplitude limitation. The PMOS-only topology of Fig. 4.13 (c) enables low power supply operation and high tuning range. The current consumption is not increased with respect to NMOS-only topologies, even quite the reverse, depending on technology (Section 4.5.3).

The implemented PMOS-only VCO circuit is presented in Fig. 4.24 (a) with a power supply voltage of 1.8V (Table 4.5). The overdrive voltage of the PMOS transistors is 0.5 V, resulting in a output DC level of 0.55 V with a power supply of 1.8 V. This means that the tuning voltage can go down to 0 V and up to 1.8 V without forward biasing the P+/N diodes. Consequently, the VCO has very wide tuning range of 28% (504MHz) (Table 4.5), without increasing the VCO gain with respect to other VCO topologies. Furthermore, the P+/N junction is never forward biased, which makes it possible to tune over the large frequency band without degrading the phase noise performance. The diodes are implemented as shown in Fig. 4.11 (a), with a Q of approximately 45 and a C_{max}/C_{min} ratio of 1.93 between 0 and 1.8V. Since the fixed capacitance at one output is 270fF due to the inductor and 550fF due to the PMOS g_m transistors and the PMOS common-source buffers, the diode capacitance is 3.43pF each at the highest frequency.

The simulated effective resistance of the total VCO is 4.7Ω. This means that 22% of the total effective resistance is due to the integrated P+/N diode varactors and the transistors, while 78% comes from the inductor. The parallel tank resistance R_p can be calculated to be near 250Ω in the higher frequency range of the VCO. The implemented transconductance of the PMOS transistors incorporates an excess loop gain of 2.5 for start-up and stable oscillation. The resulting simulated current for the total VCO is 18mA at 1.8V. The transistor sizes are shown in Fig. 4.24 (a). The simulated voltage amplitude over the LC-tank is 2.84V, which is only marginally smaller than $\frac{2}{\pi}R_p I_b$, meaning that the current through the transistors is almost a square wave. Due to the very high signal power, the calculated phase noise of the VCO is very low: -127.3 dBc/Hz at 600 kHz

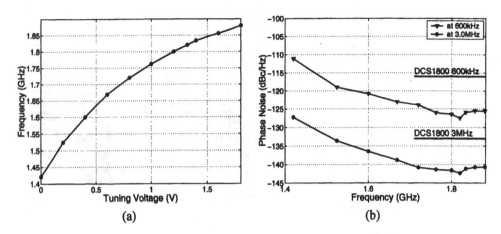

Figure 4.25: *(a) The measured tuning range between 0 and 1.8V and (b) the corresponding measured phase noise. The measured phase noise is compared to the DCS-1800 specifications (Section 2.6) at the most critical offset frequencies:* 600 kHz and 3 MHz *(the fat lines)*

and -141.3 dBc/Hz at 3 MHz offset (Table 4.5).

4.6.2.3   Experimental Results

The highly-tunable, low-phase-noise VCO is processed in a 2-metal layer 0.25μm standard CMOS technology as shown in the IC microphotograph of Fig. 4.24 (b).

Fig. 4.25 (a) shows the measured tuning range. A frequency band of 460 MHz has been measured with a maximum frequency of 1.88 GHz, revealing a tuning range of 28%. As can be seen, the tuning curve is quite linear, especially in the DCS-1800 frequency range, which is beneficial for fully integrated frequency synthesizer design.

The measured phase noise at 1.82 GHz is presented in Fig. 4.26 (a). The phase noise is as low as -127.5 dBc/Hz at 600 kHz and -142.5 dBc/Hz at 3 MHz, which is very close to the calculated values. The P+/N diodes never reach forward bias conditions, enabling the VCO to achieve the DCS-1800 noise specifications over a frequency range twice as large as required by the DCS-1800 standard, as can be seen in Fig. 4.25. Within the DCS-1800 band, the VCO exceeds the DCS-1800 specification of -116 dBc/Hz by 7.8 to 11.5 dB at 600 kHz and the specification of -133 dBc/Hz by 7.2 to 9.5 dB at 3 MHz offset (Table 4.5). Due to the high $V_{GS} - V_T$ of the PMOS transistors, the 1/f noise upconversion is fairly high, yielding a $1/f^3$ noise corner frequency of 500 kHz, as predicted in Section 4.6.1.2, in spite of the very large NMOS current source transistor (Fig. 4.24 (a)).

The circuit consumes 18mA from a 1.8V power supply, as designed. When biased at 2V and 23mA, the VCO tuning range is 30% and the measured phase noise is even -128.2 dBc/Hz at 600 kHz and -143.1 dBc/Hz at 3 MHz (Fig. 4.26 (b)). Due to the higher bias current, the $V_{GS} - V_T$ is increased and the $1/f^3$ phase noise corner becomes higher than 800 kHz. It must be stressed that these excellent results were achieved without additional processing steps, extra

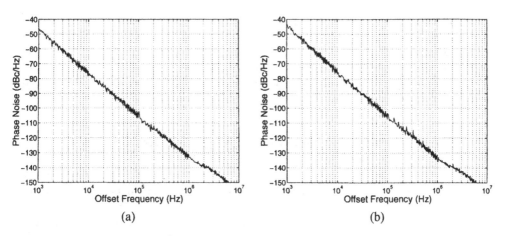

(a)   (b)

Figure 4.26: *The measured phase noise at* 1.82 GHz *and* 600 kHz *offset for (a)* 18 mA *at* 1.8V *and (a)* 23 mA *at* 2V

thick metalization, bonding wires, nor external components in a standard CMOS process with only 2 normal sized metal layers.

4.6.2.4 Quadrature Operation

The presented VCO is integrated in a fully integrated frequency synthesizer (see Chapter 5), which is implemented in a fully integrated 2V CMOS transceiver front-end for DCS-1800 [Stey00b] (see Chapter 7). For accurate up-and down-conversion, a linear quadrature oscillator signal is required, which is solved by designing the VCO such that it can be directly loaded with an optimized second order poly-phase filter, with minor performance loss and much lower transceiver power consumption.

For low phase noise and high accuracy, the combination of a differential VCO and a poly-phase filter has been chosen. Traditionally, any ASIC is a cascade of separately optimized building blocks, each designed by a specialized designer, which is beneficial for a short time-to-product. For quadrature generation, this means that the VCO and the poly-phase filter need to be buffered from each other to allow separate design. Typically, source followers are implemented to drive the poly-phase filter. However, the source followers limit the signal swing at the input of the polyphase and add non-linearity. Moreover, since typical poly-phase filter designs have a low real part for low signal loss and low noise, these buffers consume a tremendous amount of power, e.g. more than 40mW in the CMOS transceiver of [Stey98].

In the presented approach, the system is treated as a single circuit with design optimization over the building block boundaries [Stey00b] [Borr00]. For the quadrature generation, this is reflected in the direct AC coupling of the VCO and the poly-phase filter. The poly-phase filter is optimized to present a large real impedance to the VCO to limit phase noise degradation; The real input impedance is 500Ω, 400Ω of which are situated in the first stage. The resistance can not be too large for signal loss and noise. An additional advantage of the approach is that the large

VCO signal is directly fed to the poly-phase filter such that less amplification after the filter is needed. The VCO performance is only marginally affected by the direct loading; Since the extra parallel resistance is 500Ω, the phase noise is only 1.8 dB higher, which is still -125.5 dBc/Hz at 600kHz, which is more than enough for DCS-1800. The simulated power consumption is only increased marginally. The overall power saving in the full transceiver system is more than 40mW, including the smaller buffer at the output of the poly-phase filter.

4.7 Comparison with Published State-of-the-Art VCOs

In this section, the performance of the presented VCOs is compared to previously published state-of-the-art VCOs, listed in Table 4.6. For honesty of comparison, four categories are distinguished: Oscillators with external inductors (including bondwires), VCO which employ exotic techniques, such as SiGe substrates, etching techniques, ..., Si bipolar designs and fully integrated (Bi)CMOS VCOs. In Table 4.6, the state-of-the-art VCOs published in open literature with the highest performance are summarized.

The well known and commonly used Figure of Merit (FOM) of [Kin99], is used as a means to compare the different published VCOs (see Eq. (4.49)).

$$\text{FOM} = -\mathcal{L}\{\Delta\omega\} + 10\log\left[\left(\frac{\omega_0}{\Delta\omega}\right)^2 \cdot \frac{1}{P}\right] \qquad (4.49)$$

with ω_0 the VCO frequency, $\Delta\omega$ the offset, $\mathcal{L}\{\Delta\omega\}$ the phase noise at $\Delta\omega$ and P the power consumption in mW. The FOM normalizes the phase noise to frequency and offset and indicates how efficient the consumed power is used to achieve the given phase noise performance. The FOM is plotted versus the VCO output frequency in Fig. 4.27 (a). Only VCO with a FOM higher than 170 are shown. As can be seen, the presented fully integrated CMOS VCOs score high in the Figure of Merit of the integrated CMOS VCOs. Only the VCOs with filtering [Heg01, Hosh01] and VCO2 of Section 4.6.1.4 are better, but only applicable in a small frequency band. VCOs with filtering techniques are highlighted with [a] in Fig. 4.27. The DCS-1800 VCO of Section 4.6.2 still has the highest FOM of the fully integrated CMOS VCO without filtering; Only VCOs with non-standard technology tricks (such as 4μm thick top metal suspended 10μm from the substrate [Ain00]) or external tanks have a better FOM. Among the fully integrated CMOS VCOs with filtering, VCO2 is one of the best apart from the VCO in [Heg01], but this VCO profits from an extra thick top metal layer to integrate a high-quality inductor.

Note that the FOM favors high-frequency VCO design; The FOM scales with frequency (see Eq. (4.49)), but at higher frequencies, high Q inductors are more easy to implement; For a well-designed inductor, the metal losses increase less than linearly with frequency (see Fig. 4.5 (a)), such that the $Q = \omega L/R$ keeps increasing with frequency. Modern technologies have a high substrate resistance and multiple metal layers with a thick top metal. If the highest metal layers are used, the substrate losses are negligible, since the area of a high-frequency inductor is intrinsically small to allow some tuning. In addition, the small inductor leads to a small DC series resistance. The limiting factor in high-frequency VCO design is in fact the implemented varactor [DeCo01]. The conventional FOM of Eq. (4.49) relates phase noise and power consumption

	Type	ω_0 GHz	P mW	$\mathcal{L}$ dBc/Hz	$\Delta\omega$ kHz	Tuning %/V	FOM	Remarks
[Hosh01]	Bondwire[a]	0.7	2.55	-136	600	7/1.5	193.3	bias filtering
[Pfaf99]	Ext. L	1	0.625	-111	100	18/2.5	193	$Q_L > 50$
[Ain00]	BiCMOS	5.67	2.4	-119.7	1000	15/1.5	191	4μm top metal
[Heg01]	CMOS[a]	1	9.25	-135	600	16/2.5	189.8	bias filtering thick top metal
[Kuc01a]	Bondwire[a]	2	12.2	-128.5	600	35/3	188.1	cap.-switched bias filtering
VCO2	CMOS[a]	2.02	34.2	-132.5	600	11/1.8	**187.9**	bias-filtering
[Pfaf02]	Stripline	1.8	5.4	-110	100	7/1.8	187.8	PCB stripline
[Svel00]	Bondwire	1.9	2	-120.5	600	26.5/4	187.5	CMOS
[Lin00]	Bipolar	1.5	38	-148	3000	4.7/3	186.2	
[And01]	CMOS[a]	2	12.6	-140	3000	17/1.4	185.5	bias filtering off-chip
[Kuc01b]	CMOS[a]	3.6	24	-108	100	20/3	185.3	bias filtering
[Fong02]	SOI	3	3	-120	1000	58/1.4	184.8	SOI
[Kin98b]	MCM-L	2.45	5.4	-124	1000	5/2.7	184.5	CMOS
[Dec99]	Bondwire	1.9	15	-126	600	9/2.5	184.3	CMOS
[Cran95a]	Bondwire	1.8	24	-119	200	5/3	184.3	CMOS
[DeMu00a]	CMOS	1.8	32.4	-127.5	600	28/1.8	**183.1**	2-metal 0.6/1μm
[Tang99]	SOA	0.83	0.4	-100.4	100	21.6/5	182.8	SOA
[Kim00]	CMOS	0.9	20	-132	600	10/2.5	182.5	LC-ring
[DeRa01]	CMOS	17	10.5	-108	1000	8.6/1.4	182.4	
[And00]	CMOS	1.8	7.3	-121	600	11/2.7	181.9	PMOS varact.
[Wang01]	CMOS	50	13	-99	1000	2.2/2.6	181.8	
[Cran97a]	CMOS	1.8	11	-113	200	20/3	181.7	2-metal
[DeMu99a]	CMOS	2	34.2	-125.1	600	11/1.8	**180.6**	3-metal
[Yim01]	CMOS	1.8	16	-123	600	24/3.6	180.5	Switched-L
[DeCo01]	CMOS	10	50	-127	600	30/2.5	180.5	4-metal
[Zan01]	Bipolar	2.5	15.2	-104	100	20/2	180.1	Amp. Control
[Wak98]	Bipolar	1.9	21.6	-123	600	6.3/2.7	179.7	Qaudrature
[Wang98]	CMOS	9.8	11.6	-90	100	1.4/3.5	179.4	Back-gate
[Tie02]	CMOS	51	1	-85	1000	17/1.4	179.2	
[Wu00]	CMOS	10	35	-114	1000	12/2.5	178.6	distributed
[Cran97b]	CMOS	1.8	6	-116	600	12/8	177.8	2-metal
[Herz00]	CMOS	1.9	12	-100	100	14/2.5	174.8	
[Vaa01]	Bipolar	4.3	36.5	-113	600	46/2.5	174.5	LC-relax.
[Kin98a]	CMOS	4.7	11	-90	100	4/2.7	173	

[a] indicates the use of a filtering technique

able 4.6: *An overview of the published state-of-the-art VCOs in all sorts of technology with the ighest FOM.*

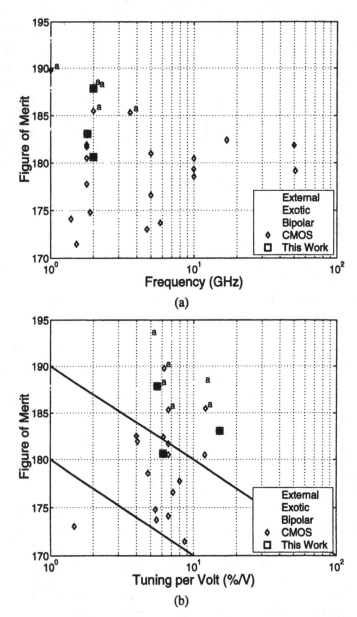

Figure 4.27: (a) *The conventional FOM versus the VCO output frequency for different kinds of VCOs. (b) The conventional FOM normalized to the input of the VCO. The lines denote a constant FOM-tuning (per volt) product. The [a] means that a filtering technique is applied in that particular VCO.*

ut it does not include one of the important VCO specifications: the tuning range. Nevertheless, the tuning is related to power and phase noise. A high tuning at low power supply voltage requires a high VCO gain, which makes the VCO much more sensitive to voltage induced phase noise. The extra tuning devices will further degrade the noise performance due to their non-zero resistance. Obviously, wide tuning range and low phase noise are contradictory demands. Therefore the figure of merit is plotted against the ratio of the tuning range and power supply voltage in Fig. 4.27 (b). This ratio is a measure of the VCO gain, and thus of the noise sensitivity.

The lines in Fig. 4.27 (b) indicate constant FOM-tuning per volt products, following Eq. (4.50):

$$\text{FOM}_T = -\mathcal{L}\{\Delta\omega\} + 10\log\left[\left(\frac{\omega_0}{\Delta\omega}\right)^2 \cdot \frac{1}{P}\right] + 10\log\left(\frac{\text{Tuning}}{V}\right) \qquad (4.50)$$

The last term is the ratio of the tuning range in % and the power supply voltage and divides the normalized phase noise by the VCO gain. In fact, the normalized phase noise is referred to the input of the voltage controlled oscillator. The lower the input referred phase noise, the better the VCO performance. The presented VCOs outperform most published fully integrated CMOS VCOs in Table 4.6, taking into account all important VCO specification: phase noise, tuning range, frequency and power (see Fig. 4.27 (b)). Again, the filtered VCOs (highlighted with [a]) perform better but are only applicable in a small frequency band. Even compared to the filtered VCOs, the DCS-1800 VCO has a comparable FOM-tuning product, without using filtering techniques and without extra thick metal layers, etching, ... in a standard 0.25μm CMOS technology with 2 normal sized metal layers.

4.8 Conclusion

In this chapter, the design of fully integrated, low-phase-noise VCOs for high-quality communication systems is discussed. The oscillator choice is restricted to passive LC-tank VCOs, since relaxation and ring oscillators and active inductor implementations, despite their excellent integratability, require intrinsically a high power for low phase noise.

Since phase noise is critical in communication systems, a design oriented non-linear phase noise theory is presented; In contrast to linear phase noise theories, the non-linearity of the active element has been taken into account. The phase noise theory uncovers the mechanisms of noise upconversion without going into tedious theoretical derivations. In addition, the distinction has been made between amplitude and phase noise. The resulting phase noise formula provides insight on how the active circuit design influences the VCO phase noise by definition of the *excess loop gain*. For excess loop gains higher than 2, amplitude noise has been proven to be negligible and the phase noise has been related to the excess loop gain. The formula is a worst-case analysis by assuming a square wave noise transfer; Actual circuit implementations provide a smoother transfer and thus less noise upconversion. The presented theory relates in a single formula, the oscillator phase noise to the non-linear active circuit and thus noise upconversion by definition to a designer defined parameter, the excess loop gain, and therefore enables a more optimized VCO design.

From the phase noise analysis, the importance of the quality of the LC-tank emerged; The effective series resistance of the tank should be as low as possible (power of 3 in noise) and the inductance needs to be as high as possible (power of 4 in noise); In fact, oscillator inductor design comes down to optimizing the Q, with the stress on a high inductance. For mass production, fully integrated spiral inductors, without etching or extra thick metals provide almost perfect yield and no extra processing cost. However their losses are high in standard CMOS technology; Losses in the metal are due to the DC metal resistance and high-frequency effects such as the skin-effect and eddy-currents. Guidelines to lower these losses are not to use wide metal tracks and realize "hollow" inductors. Losses in the substrate are due to substrate currents induced by the magnetic field of the inductor. Guidelines to lower these losses are a high substrate resistivity (in processing or patterned ground shields) and limitation of the inductor area (only for highly-conductive substrates). Since the electromagnetic interactions in integrated inductors are far too complex to analyze, an inductor simulator-optimizer program has been developed; The program implements the inductor simulator FastHenry in a simulated annealing loop resulting in an optimized inductor geometry for a given set of technology and circuit parameters and specifications. The program is deployed to design balanced octagonal inductors; Octagonal to approximate the circular inductors and balanced for the higher area-Q efficiency. In addition, varactors contribute to the effective resistance; For CMOS implementation, P+/N diodes are identified as the best solution since they result a rather constant VCO gain (important for PLL implementations) in combination with a high and constant Q. In view of CMOS technology scaling, accumulation-mode MOS varactors become a more efficient solution, since their Q scales with the square of the gate length, while providing a high capacitance ratio.

Next, general VCO circuit design is discussed; CMOS has been identified as a more powerful technology than bipolar for high-frequency, low-phase-noise VCOs. The operating frequency of bipolar VCOs is limited by the base resistance and the much higher $\frac{g_m}{I}$ ratio results in a low current, which significantly reduces the VCO signal power and thus the phase noise. Only CMOS balanced circuit topologies are elaborated; For high power supply voltages (> 2V), the complementary topology presents the most power-efficient solution, since its amplitude is twice as high for the same current. For lower power supply voltages, xMOS-only topologies provide lower overall phase noise, but at the expense of higher power. Depending on the technology PMOS-only or NMOS-only are more power efficient. Finally, $1/f^3$ phase noise mechanisms are elaborated; The noise from the bias current source has been identified as the main source of flicker phase noise if any imbalance exists in the circuit. The noise is modulated onto the carrier by even harmonics present in the oscillator signal. An easy to use method to measure the sensitivity of a VCO circuit to $1/f$ noise upconversion is presented, based on the derivatives of the signal slopes. A flicker noise corner factor γ_{1/f^3} is defined, which is easily extracted from HSpice simulations.

Last but not least, all the prior theory is applied to design high-frequency, low-phase-noise VCOs in standard CMOS technology. The first implementation is a 2 GHz VCO set with flicker noise upconversion minimization. The flicker noise corner factor γ_{1/f^3}-method is used to determine the optimal bias point of the VCOs and is used to explore circuit techniques to minimize $1/f$ noise upconversion. The phase noise of the VCOs as low as -125 dBc/Hz at 600kHz with a $1/f^3$ phase noise corner of less than 15 kHz in a standard 0.65μm CMOS process. In a small frequency band, the phase noise of the VCO is as low as -132.5 dBc/Hz at 600kHz with no flicker noise at all. This is accomplished through bias current source filtering, confirming the contribution of bias current noise in unbalanced designs. A second implementation is a fully integrated VCO with a tuning range of 28% and a phase noise of less than -127.5 dBc/Hz at 600kHz, which exceeds the severe DCS-1800 specifications using only a 2-metal, standard $0.25\mu m$ CMOS technology. Additionally, the VCO is designed to drive a poly-phase filter for accurate quadrature outputs without additional buffering. To show the quality of the presented designs, the VCOs are compared to existing state-of-the-art VCOs; The DCS-1800 VCO in $0.25\mu m$ CMOS proved to be one of the best fully integrated CMOS VCOs, according to the FOM combining phase noise, power and tuning range. Only VCOs with external inductor, exotic techniques or bias current source filtering perform better. The presented VCO with flicker noise minimization by bias filtering is one of the best fully integrated CMOS VCO published including the ones with bias filtering, without reverting to non-standard techniques, such as etching, extra thick top metal layers as in [Heg01], thick field oxide or special substrate doping.

Chapter 5

Monolithic Phase-Locked Loops

5.1 Introduction

The feasibility of monolithic integration of a low-phase-noise VCO and a high-speed prescalers has been proven in the previous chapters; Moreover, the VCO is capable of driving a polyphase filter without the need for additional buffers, such that accurate quadrature LO signals can be provided to the receive and transmit path in a power efficient manner. These realizations pave the way to the integration of a complete phase-locked loop in CMOS, without trimming, tuning and external components. To demonstrate that high-quality PLL frequency synthesizers are feasible in standard CMOS technology, the stringent DCS-1800 class I/II communication standard has been chosen as driving application for this design. For clarity, the DCS-1800 synthesizer specification summary of Table 2.2 is repeated in Table 5.1 with addition of the impact of every PLL building block on the spec.

The 16-modulus prescaler of Section 3.5 is a custom design for the DCS-1800 frequency synthesizer. The division moduli from 65 to 73 are necessary to synthesize the full DCS-1800 frequency range with a 26 MHz reference Xtal oscillator (row 7 in Table 5.1). The low-phase-noise VCO of Section 4.6.2 is integrated in the PLL in a plug 'n' play fashion. The tuning range of over 30% enables the synthesis of the DCS-1800 frequency range over process variations without the need for an integrated capacitance bank. The VCO determines the out-of-band phase noise of the PLL and must exceed the noise spec at 600kHz and 3MHz offset (row 1 in Table 5.1).

One obstacle remains: the integration of the loop filter. The goal of the design is twofold: To minimize the integrated capacitance and exceed the phase noise (at least -123 dBc/Hz at 600kHz offset) and dynamic performance (at most $300\mu s$) required by DCS-1800 (see Table 5.1) to show the feasibility of high-quality monolithic PLLs in CMOS. The main trend in frequency synthesizer design is to go towards higher loop bandwidths which is beneficial for integration, for the loop dynamics and for the rms phase error. However, the noise of the filter resistors and active elements tends to drown the VCO noise at the critical offset frequencies, such as 600 kHz (row 1 Table 5.1), nullifying the effort to integrate a high-quality VCO. To guarantee the phase noise performance under settling, rms phase error and area constraints, an intermediate filter bandwidth of 35kHz has been chosen, as elaborated in Section 5.3.3. The filter order is three

Phase Noise		
at 600 kHz offset	-116 dBc /Hz	Loop filter/VCO (-127.5 dBc /Hz)
at 3 MHz offset	-133 dBc /Hz	VCO (-142.5 dBc /Hz)
Spurious Suppression		
Reference Spurs	< -80 dBc	Loop filter/Charge Pump
Fractional Spurs	see Fig. 2.24	Chapter 6
rms Phase Error $\Delta \Phi_{rms}$	2^0	Loop filter/Charge pump
Settling Time (95 MHz /180 Hz)		
DCS-1800	865 μs	Loop Filter (C_{tot})
(E)GPRS	310 μs	Charge Pump (I_{qp})
(E)(HS)CDS	288 μs	
Frequency Resolution	200 kHz	Chapter 6
Center Frequency	1.8 GHz	VCO/Prescaler
Tuning range	170 MHz	VCO (30%)
		Prescaler ($N = 65..79$)
Power Consumption	*ALAP*	All
Area	*ALAP*	All
Supply Voltage	1.8V-2V	All

Table 5.1: *Summary of the frequency synthesizer specifications in a DCS-1800 communication system and the impact of the different PLL building blocks.*

to provide sufficient suppression of phase noise generated by the passive filter components (and of the 3rd-order $\Delta\Sigma$ modulator in Chapter 6) and of spurious tones at the reference frequency. Lower filter orders lead to high phase noise at 600 kHz and higher filter orders introduce too much capacitance to maintain the noise spec of a monolithic solution. To minimize the integrated capacitance further, a dual-path loop filter topology (Section 5.3.1) has been developed which reduces the capacitance with a factor 4 over a conventional topology for the same phase noise.

The focus of this chapter is on the filter topology selection and optimization for minimal capacitance with maximum phase noise performance. The circuit implementation of the filter and the equalizer, which equalizes the loop gain over the full frequency range, are discussed next. The monolithic 4th-order, type-II PLL is integrated in a standard 0.25μm CMOS process with only two metal layers. The experimental results are presented which show full compliance with the DCS-1800 class I/II specifications while the occupied chip area is only 2x2 mm^2. The resulting phase noise performance of the monolithic PLL is even better than that of most integrated VCOs!

5.2   Loop Filter Topology Selection

The loop filter choice is critical for the specifications of the PLL. In Section 2.4.2, possible loop implementations up to a 3rd-order, type-II loop with their corresponding performance are elaborately discussed. To implement a type-II PLL, an active filter implementation is necessary for the integration in the filter transfer function. However, a way exists to implement a type-II passive loop: the charge pump PLL.

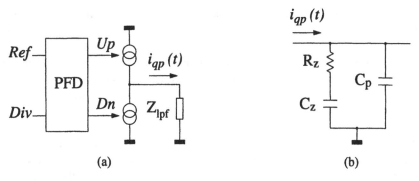

Figure 5.1: *(a) Phase-frequency detector and charge pump and (b) a loop filter for a type-II, third-order charge pump PLL.*

5.2.1 Charge Pump PLL

The combination of a phase-frequency detector (PFD) (see Section 2.4.1.4) and a charge pump Fig. 5.1 (a) is widely used in PLLs [Egan81], because of its frequency-sensitive error signal, that can aid acquisition when the loop is out of lock. If the pump currents of the charge pump are both I_{qp}, the phase detector gain is

$$K_{pd} = \frac{I_{qp}}{2\pi} \tag{5.1}$$

A loop filter implementation for a charge pump PLL is shown in Fig. 5.1 (b). The open loop transfer function becomes

$$GH(s) = \frac{I_{qp}K_{vco}}{2\pi N} \cdot \frac{G_{lpf}(s)}{s} \tag{5.2}$$

$$G_{lpf}(s) = \frac{1 + s\tau_z}{s\left(C_z + C_p\right)\left(1 + s\tau_p\right)} \tag{5.3}$$

with $\tau_z = R_z C_z$ and $\tau_p = R_z \cdot \left(C_z^{-1} + C_p^{-1}\right)^{-1}$.

This is the transfer function of a third-order, type-II PLL as in Fig. 2.13. There are two poles at zero frequency; One caused by the VCO and one caused by the charge pump under lock condition, which constitutes an ideal integrator. A charge pump PLL is the only way to realize a type-II behavior with only a passive filter. The crossover frequency is approximately

$$\omega_c \approx \frac{I_{qp}K_{vco}}{2\pi N} \cdot \frac{R_z C_z}{C_z + C_p} \tag{5.4}$$

The placement of the pole is a trade-off between the necessary spurious tone and phase noise suppression and the dynamic behavior of the loop. As a rule of the thumb, the zero is placed at one fourth of the crossover frequency and the pole at four times the crossover frequency. This placement guarantees sufficient phase margin (around 60°) for loop stability and sufficient damping to suppress ringing.

To ensure that the loop can track even very small phase errors, it is advised to use a zero-dead-zone PFD [Gard79]. In the zero-dead-zone PFD circuit, the charge pump current sources always conduct current, even when the loop is in lock. This means that the charge pump never is in its high-impedance state and the loop is always closed under lock condition. When a dead zone is present, the phase noise is not suppressed at very low offset frequencies proportional to the size of the dead zone. The zero-dead-zone PFD is discussed in more detail in Chapter 6. Another important issue in PFD-charge pump design, is the minimization of spur generation, which is again elaborated in Chapter 6 .

To check whether the type-II, third-order charge pump PLL can be used for integration in CMOS wireless transceiver, the phase noise contributions of the charge pumps and the loop filter are calculated. The noise of the two currents sources of the charge pump when the loop is in lock can be written as

$$di_{qp}^2 = 2\alpha_{qp} \cdot di_n^2 \tag{5.5}$$

with α_{qp} the time fraction that the current sources are active. In this book, the on-time of the charge pumps in lock is approximately 7.5% of one reference cycle. This value is a trade-off between the minimization of the dead zone and the sensitivity to spurious tones; Whenever, the charge pumps are on, noise coupling through the substrate and the power supply lines can introduce spurious tones in the PLL output spectrum. Therefore, the on-time of the pumps must not be taken too long, but long enough to ensure the removal of the dead zone. The transfer function for this noise source is:

$$T_{qp,o}(s) = \frac{\theta_{out}}{i_{qp}} = \frac{G_{lpf}(s) K_{vco} N}{sN + K_{pd} G_{lpf}(s) K_{vco}} \tag{5.6}$$

Knowing that the noise current of current source is $di_n^2 = 4kT g_{m,n} \, df$ with k the Boltzmann constant, T the temperature and $g_{m,n}$ the transconductance of the current source transistor, the phase noise can be calculated at a $\Delta\omega$ offset:

$$\mathcal{L}_{qp}\{\Delta\omega\} = \frac{\theta_{out}^2(\Delta\omega)}{2} \tag{5.7}$$

$$= \left| T_{qp}(\Delta\omega) \right|^2 \cdot \alpha_{qp} \cdot 4kT \frac{2I_{qp}}{(V_{GS} - V_T)_{qp}} \tag{5.8}$$

The noise is now calculated for a charge pump current of 10 μA, an on-time fraction α_{qp} of 7.5% and an overdrive voltage $(V_{GS} - V_T)_{qp}$ of 0.3 V. The bandwidth is chosen $\omega_c = 2\pi \cdot 70$ kHz for integratability and the lowest DCS-1800 frequency is taken as a worst case situation with $N=65$ and a $K_{vco} = 2\pi \cdot 300$ MHz/V. The passive elements can be calculated to be $R_z = 10.1$ kΩ, $C_z = 895$ pF and $C_p = 60$ pF. The resulting phase noise at 600 kHz is $\mathcal{L}_{qp}\{600\text{kHz}\} = -124.3$ dBc/Hz for a total capacitance of around 1 nF. The noise spec is below the DCS-1800 phase noise specification (Section 2.6), but much higher than the VCO noise, leaving not much margin for the noise contributions of the other building blocks. The charge pump current noise contribution can be lowered by increasing the overdrive voltage or decreasing the on-time, but both action tend to deteriorate the spurious suppression. The same calculation can be performed for the thermal noise from R_z: $di_{R_z}^2 = 4kT/R_z \, df$. The transfer function is:

$$\frac{\theta_{out}}{i_{R_z}} = \frac{K_{vco}N}{sN + G_{lpf}(s)K_{pd}K_{vco}} \cdot \frac{R_z C_z}{sR_z C_z C_p + (C_z + C_p)} \tag{5.9}$$

The first part is the transfer function from the output of the filter to the PLL output and the second factor is the transfer function of the resistor current noise to voltage filter output noise. The phase noise at 600 kHz can be calculated in the same way as above. $\mathcal{L}_{R_z}\{600 \text{ kHz}\} = -114.4$ dBc/Hz, making it impossible to achieve the DCS-1800 specification. To reduce the phase noise, the loop bandwidth must be lowered, causing an increase of integrated capacitance. For example, to lower the noise to -125 dBc/Hz, which together with the VCO noise gives the wanted -123 dBc/Hz, the loop bandwidth must be lowered by a factor 2.33. The resulting increase in integrated capacitance is a factor 2.33^2 or from 955 pF to almost 5.2 nF, which is too large to integrate. Using the first order approximation of Eq. (2.33), the settling time is also increased by a factor 2.33.

5.2.2 Fourth-Order PLL

To reduce the phase noise contributions of the charge pump and the loop filter resistor R_z, an extra pole can be added at the same frequency or a little further than ω_p. This further suppresses the phase noise at high offset frequencies and might allow to relax the sizing of the loop parameters. To implement the fourth-order, type-II charge pump PLL, an active loop filter implementation is chosen.

When using only passive components, some problems arise in the charge pump. In the configuration of Fig. 5.1, the output voltage of the pump varies with the VCO control voltage. As a result, the overdrive voltages of the switches and the current sources will not be the same for the top source and the bottom source of the charge pump. The resulting mismatches in up- and down currents cause a net charge injection in the loop filter, even when the loop is locked and thus reference frequency spurs. Another problem is that a large output voltage range is required to allow VCO tuning without loosing the saturation condition in the transistors. But as deducted in Eq. (5.8), the $(V_{GS} - V_T)_{qp}$ must be high for low current noise. So despite the charge pump PLL design, an active filter will have to be used to set the charge pump output to a fixed level. A straightforward implementation is shown in Fig. 5.2, together with the bode plot of its open loop gain and phase. The extra pole ω_4 is placed on top of ω_p, by making $R_4 C_4 = \tau_p$. The resistor R_4 introduces an extra noise source, but this noise contribution can be made arbitrarily small by increasing C_4. To ensure sufficient phase margin the placing of both ω_4 and ω_p must be altered with respect to the third-order loop. Both poles must be placed six times higher than the crossover frequency instead of four times. The high frequency phase noise and spurious suppression is increased by a factor $(6/4)^2(6\omega_c/\Delta\omega)^2$ in comparison with the 3rd-order charge pump PLL; To make the phase noise contribution of R_4 smaller than that of R_z, C_4 must be chosen large, so that the total integrated capacitance once again becomes too large (see next section).

In addition, the noise contribution of the active part of the filter must be taken into account. Since the noise of R_z is filtered by the additional pole, the noise of R_4 and of the opamp become dominant. To make sure that the phase noise contribution of the opamp is below -125 dBc/Hz

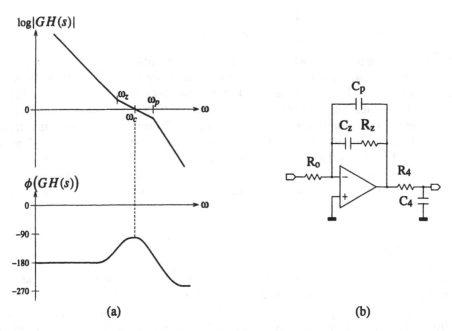

Figure 5.2: *(a) Open loop gain and phase of a type-II, 4th-order charge pump PLL (b) an active loop filter implementation.*

at 600 kHz offset, the G_{mA} must be higher than 0.6 mS, which poses no real threat to the power consumption. R_4 is chosen a factor 3 smaller than R_z for low noise. The calculations are performed for the same loop parameters as in the previous section for $\omega_c = 2\pi \cdot 30$ kHz revealing a total capacitance of 5.82nF for a total phase noise of merely -121 dBc/Hz at 600 kHz. This is an unacceptable result considering that the VCO noise is as low as -127 dBc/Hz.

5.3 Dual-Path Fourth-Order PLL

It is shown in the previous chapter that it is difficult to realize the phase noise specifications imposed by the DCS-1800 standard, without the need for several nF of integrated capacitance. A fourth pole helps to reduce the phase noise, but the best remedy remains the reduction of the loop bandwidth. When the loop bandwidth is reduced, the deceleration of the loop's dynamic behavior and the explosive increase in capacitance must be attacked. A solution to integrate a relatively low-bandwidth, higher-order PLL with limited capacitance is the dual-path filter topology developed in [Miju94, Cran98]. The next sections discuss the topology implementation and optimization of a dual-path loop filter that achieves the DCS-1800 specifications.

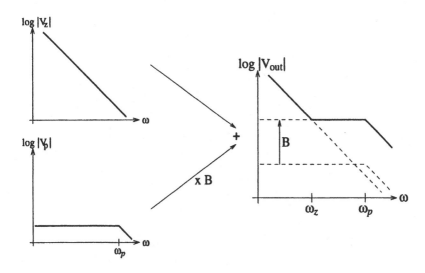

Figure 5.3: *The dual-path loop filter principle. The low frequency zero ω_z is realized by adding both filter paths.*

5.3.1 Dual-Path Filter Topology

The principle of the dual-path filter topology is explained in Fig. 5.3. In a conventional passive filter of a third-order charge pump PLL (Fig. 5.1 (b)), the capacitor C_z that realizes the compensating zero is the most area-expensive component. The dual-path filter topology circumvents the need to physically integrate the zero with an RC combination at the desired frequency. By combining two signal paths, a virtual zero is realized. The placement of the zero is controlled by the relative amplification between both paths.

The first signal path V_z is an integration of the input current and the second is a low pass filter V_p, with a pole at ω_p.

$$V_z = \frac{1}{sC_z} \cdot I_{in} \tag{5.10}$$

$$V_p = \frac{R_p}{1 + sR_pC_p} \cdot I_{in} \tag{5.11}$$

These two signal paths are now combined to form the virtual zero, with a relative amplification by a factor B.

$$
\begin{aligned}
V_{out} &= V_z + B \cdot V_p \\
&= \left[\frac{1}{sC_z} + B \cdot \frac{R_p}{1 + sR_pC_p} \right] \cdot I_{in} \\
&= \frac{1}{sC_z} \cdot \frac{1 + s\tau_z}{1 + s\tau_p} \cdot I_{in}
\end{aligned}
\tag{5.12}
$$

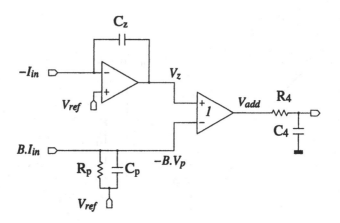

Figure 5.4: *The dual-path loop filter implementation.*

with $\tau_z = R_z \cdot (C_p + B \cdot C_z) \cong BR_pC_z$ and $\tau_p = R_pC_p$.

A low-frequency zero is realized, without the need for a large integrated capacitor, because τ_z is multiplied by a factor B. The dual-path filter can be implemented with an additional pole ω_4 as shown in Fig. 5.4. Note that the number of capacitors is the same as for the conventional 4th-order, type-II implementation of Fig. 5.2 (b), but it is shown in Section 5.3.3 that the total capacitance value is much lower. The circuit has two current inputs; one is integrated to form V_z, the other is low-passed to V_p. The input currents are delivered by two charge pumps with identical switch control signals. The implementation of the multiplication factor B is done in the current domain by using a pump current that is a factor B larger for the charge pump driving the signal V_p. The integration in the top path is done actively, to make sure that the charge pump output is biased at V_{ref} for spurious reduction. The bottom path can be implemented passively, because no current will run through R_p, when the loop is in lock, holding the charge pump output at V_{ref}. The circuit implementation is discussed in Section 5.4.2.

5.3.2 Transfer Functions

5.3.2.1 Open Loop Gain

The loop filter impedance of Fig. 5.4 can be calculated as follows using Eq. (5.12):

$$V_{out} = (V_z + B \cdot V_p) \cdot \frac{R_4}{1 + sR_4C_4}$$

$$= \frac{1}{sC_z} \cdot \frac{1 + s\tau_z}{(1 + s\tau_p)(1 + s\tau_4)} \cdot I_{in} \qquad (5.13)$$

with $\tau_4 = R_4C_4$ and $\tau_z = R_p \cdot (C_p + B \cdot C_z) \cong BR_pC_z$ and $\tau_p = R_pC_p$. The open loop gain is than:

$$GH(s) = \frac{I_{qp}K_{vco}}{2\pi N} \cdot \frac{1 + s\tau_z}{s^2C_z(1 + s\tau_p)(1 + s\tau_4)} \qquad (5.14)$$

with a crossover frequency :

$$\omega_c = \frac{I_{qp} K_{vco}}{2\pi N} \cdot \frac{(C_p + BC_z)R_p}{C_z}$$ (5.15)

To ensure sufficient phase margin, the zero $\omega_z = 1/\tau_z$ is placed a factor α below the crossover frequency. The two high-frequency poles $1/\tau_p$ and $1/\tau_4$ are chosen to coincide in order to have the best trade-off between out-of-band noise suppression and phase margin. They are placed a factor β above the crossover frequency. An extra degree of freedom can be introduced by defining a factor γ. If R_4 is made γ times smaller than R_p, C_4 must be made γ times larger than C_p. This enables the trade-off between phase noise and integrate capacitance. The passive filter elements can now be calculated using the following equations:

$$R_p = \frac{2\pi N}{I_{qp} K_{vco}} \cdot \frac{\omega_c}{B} \qquad\qquad C_4 = \gamma \cdot C_p$$

$$C_p = \frac{I_{qp} K_{vco}}{2\pi N} \cdot \frac{B}{\beta \omega_c^2} \qquad\qquad R_4 = \frac{R_p}{\gamma} \qquad\qquad (5.16)$$

$$C_z = \frac{I_{qp} K_{vco}}{2\pi N} \cdot \frac{\alpha}{\omega_c^2}$$

From the open loop gain, the transfer function for the VCO noise and the Xtal reference noise can be calculated straightforwardly using the LTI PLL model.

$$T_{vco,o} = \frac{\theta_{out}}{\theta_{vco}} = \frac{s^2 C_z N (1 + s\tau_p)(1 + s\tau_4)}{s^2 C_z N (1 + s\tau_p)(1 + s\tau_4) + K_{pd} K_{vco}(1 + s\tau_z)}$$

(5.17)

$$T_{ref,o} = \frac{\theta_{out}}{\theta_{ref}} = \frac{N K_{pd} K_{vco}(1 + s\tau_z)}{s^2 C_z N (1 + s\tau_p)(1 + s\tau_4) + K_{pd} K_{vco}(1 + s\tau_z)}$$ (5.18)

5.3.2.2 Charge Pump Noise

The noise contribution of both charge pumps to the output phase noise can again be calculated. The transfer function for both pumps is given by:

$$T_{qp,z,o} = \frac{\theta_{out}}{i_{qp,z}} = \frac{s C_z N K_{vco}(1 + s\tau_p)}{s^2 C_z N (1 + s\tau_p)(1 + s\tau_4) + K_{pd} K_{vco}(1 + s\tau_z)} \cdot \frac{1}{s C_z}$$ (5.19)

$$T_{qp,p,o} = \frac{\theta_{out}}{i_{qp,p}} = \frac{s C_z N K_{vco}(1 + s\tau_p)}{s^2 C_z N (1 + s\tau_p)(1 + s\tau_4) + K_{pd} K_{vco}(1 + s\tau_z)} \cdot \frac{R_p}{1 + s\tau_p}$$ (5.20)

with the noise magnitudes given by $di_{qp}^2 = 2\alpha_{qp} \cdot di_n^2$ with an on-time $\alpha_{qp} = 0.075$ and $di_{qp,p}^2 = B di_{iq,z}^2$. The noise current of current source is

	g_m	KF	AF	W	L	C_{ox}
NMOS	5.5 μS	6.71e-25 V^{2}F	0.849624	25μm	2.2μm	5.8e-3 F/m^2
PMOS	7.7 μS	7.94e-26 V^{2}F	0.96485	7μm	10 μm	5.12e-3 F/m^2

Table 5.2: *Values of the different noise parameters in the 0.25μm CMOS technology*

$$di_n^2 = 4kT g_m \, df + \frac{KF_n \cdot g_{m,n}^2}{W_n L_n C_{ox} \cdot f^{AF_n}} + \frac{KF_p \cdot g_{m,p}^2}{W_p L_p C_{ox} \cdot f^{AF_p}} \qquad (5.21)$$

with g_m the transconductance of the current source transistors, W and L the transistor sizes, C_{ox} the gate oxide capacitance and KF and AF 1/f noise transistor model parameters. The values of the different parameters for the 0.25μm CMOS technology are listed in Table 5.2. The 1/f noise of the current source transistors is incorporated since the transistors are usually quite small due to the small charge pump current ($I_{qp} = 1.3\mu$A); The resulting 1/f noise could seriously degrade the in-band noise of the PLL.

An additional advantage of the dual-path loop filter shows up, when the charge pump current transfer functions of the conventional and the dual-path 4th-order loop are compared; It can be shown that the phase noise contribution of the charge pumps is a factor B smaller for the dual-path case for similar loop parameters. The noise of the multiplied pump $I_{qp,p}$ is dominant by a factor $\alpha\beta$, since the ratio of the transfer functions is proportional to $C_z/C_p = \alpha\beta/B$ and the current is B times larger.

5.3.2.3 Loop Filter Noise

The noise transfer functions for the different loop filter current noise sources are calculated in the same way as above.

$$T_{R_p,o} = \frac{\theta_{out}}{i_{R_p}} = \frac{sC_z N K_{vco}(1 + s\tau_p)}{s^2 C_z N(1 + s\tau_p)(1 + s\tau_4) + K_{pd} K_{vco}(1 + s\tau_z)} \cdot \frac{R_p}{1 + s\tau_p} \qquad (5.22)$$

In the conventional loop filter, R_z was critical for noise. Since R_z is here replaced with BR_p, the phase noise contribution of R_p is also a factor B smaller with respect to the contribution of R_z. The same applies to the noise contribution of R_4.

$$T_{R_4,o} = \frac{\theta_{out}}{i_{R_4}} = \frac{sC_z N K_{vco}(1 + s\tau_p)(1 + s\tau_4)}{s^2 C_z N(1 + s\tau_p)(1 + s\tau_4) + K_{pd} K_{vco}(1 + s\tau_z)} \cdot \frac{R_4}{sC_4(Z_a + R_4) + 1} \qquad (5.23)$$

Z_a is the output impedance of the loop filter adder. Noise simulations in the next section reveal that especially R_4 can become critical at higher offset frequencies, since its noise is an order less filter than that of R_p.

Two active elements are present in the loop filter, both generating noise. Their noise sources are approximated by the equivalent input noise voltage sources $v_{gm} = 4kT/g_m$. Again, the 1/f noise of the active elements is taken into account. The 1/f noise is minimized by using PMOS biasing and large transistors. The shapes of the noise transfer functions of the integrator opamp and the adder are identical.

$$T_{g_m,o} = \frac{\theta_{out}}{v_{g_m}} = \frac{sC_zNK_{vco}(1+s\tau_p)}{s^2C_zN(1+s\tau_p)(1+s\tau_4) + K_{pd}K_{vco}(1+s\tau_z)} \qquad (5.24)$$

The g_m of both the adder and the integrator amplifier allow a power-noise trade-off. As long as the g_m is higher than 1mS, the actively generated noise in the filter is not critical.

5.3.3 Filter Optimization

To start the optimization of the loop filter, the fixed parameters of the PLL are determined. In this application, the reference frequency is set to 26 MHz. The parameters of the VCO are known (see Section 4.6.2); The VCO gain K_{vco} is between $2\pi \cdot 300$ MHz/V (for $N = 65$) and $2\pi \cdot 120$ MHz/V ($N = 72$) in the desired frequency range. The phase noise of the VCO is -127.5 dBc/Hz at 600 kHz and -142.5 dBc/Hz at 3 MHz offset from a 1.82 GHz carrier. The optimization is performed in Matlab [Mat97], using Eq. (5.16) to determine the loop components and the total integrated capacitance. The noise transfer functions of Eq. (5.18)-Eq. (5.24) are used to evaluate the phase noise contributions, using the relationship $\mathcal{L}\{\Delta\omega\} = \theta_{out}^2(\Delta\omega)/2$. The model allows to rapidly explore the effect of the loop parameters on the overall PLL performance.

The loop parameters that are incorporated in the optimization process are the loop bandwidth ω_c, the charge pump current I_{qp}, the charge pump current factor B and the fourth pole factor γ. The factors α and β in Eq. (5.16) are no optimization parameters, because they set the phase margin and thus the loop stability. The chosen values are $\alpha = 4$ and $\beta = 6$. The optimization goal is to achieve the phase noise specification within the following constraints. The first constraint is low power consumption, mainly reflected in the choice of the g_m of the adder and the integrator amplifier .The second constraint is the occupied chip area; The total integrated capacitance, which is the sum of C_z, C_p and C_4, must have a value that allows integration in silicon, i.e. < 2nF. The phase noise constraint is set at -123 dBc/Hz at 600kHz; Although the phase noise specification of DCS-1800 is tougher at 3MHz, 600kHz has been chosen, since the passive loop parameters can degrade the phase noise at this offset frequency, while at 3MHz the noise is determined by the VCO. The last constraint is the rms phase error which must be lower than 2° (see Table 2.2).

The charge pump current I_{qp} and the loop bandwidth ω_c are the key design parameters which set the necessary capacitance, the phase noise and rms phase error. To explore the design space, simulations are performed for different combinations of currents and bandwidths at $N = 72$. The results are plotted in Fig. 5.5 for $B = 12$ and $\gamma = 3$. Note that the charge pump current is the basic current; The equalization in the PLL is also incorporated in the simulation, which means that for lower VCO gains (higher N), the charge pump current is increased to keep the open loop gain constant. The basic charge pump current is limited by the parasitic charge injection of the switching transistors in the charge pump. For currents smaller than $0.5\mu A$, the parasitic charge injection becomes more and more dominant, giving rise to increased spurs at the reference frequency. In Fig. 5.5 (a), the capacitance is plotted in nF. The shaded area indicates capacitance values higher than 10nF; In this area, the capacitance increases almost exponentially and is

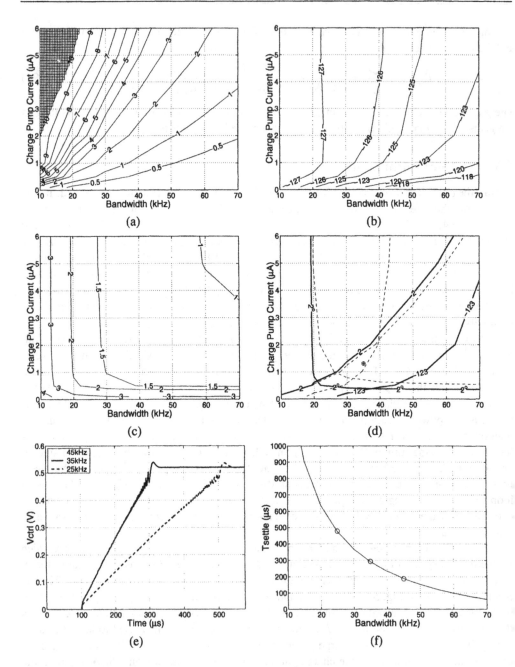

Figure 5.5: *Matlab simulation results: (a) The integrated capacitance (nF), (b) the phase noise at 600 kHz (dBc/Hz), (c) rms phase error and (d) the demarcation of the design space for N=72 (solid) and N=65 (dashed). The time-domain (e) and hyperbolically extrapolated (f) loop dynamics for different bandwidths.*

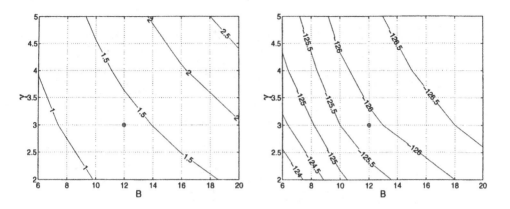

Figure 5.6: *The influence of the charge pump multiplication factor B and the 4th pole factor γ on (a) the integrated capacitance (nF) and (b) the total output phase noise (dBc/Hz at 600kHz)*

therefore forbidden territory. As could be expected, higher bandwidths and lower charge pump currents result in less capacitance. The phase noise, shown in Fig. 5.5 (b), is within spec due to the low K_{vco} for $N = 72$. For the rms phase error, the bandwidth needs to be higher than 25kHz for all charge pump currents within the useful range (see Fig. 5.5 (c)); To compute the rms phase error, the noise bandwidth has been determined to be $f_{nbw} \approx 1.16 f_c$. In Fig. 5.5 (d), the wanted spec lines are brought together to demarcate the design space. The solid lines are for $N = 72$ and show a design space that favors high bandwidths, due to the low K_{vco}. However at the lower end of the frequency range of the PLL for $N = 65$, the story is totally different; The VCO gain K_{vco} is much higher, making the phase noise specification at 600kHz much more stringent, as indicated in Fig. 5.5 (d) with the dashed lines. The design space shrivels and the bandwidth needs to be restricted to maintain the wanted phase noise specification. The bandwidth is set to 35kHz and the charge pump current to 1.3μA (see marker in Fig. 5.5 (d)). The marker is on the lower end of the design space, to keep the amount of capacitance low (≈ 1.4nF).

The settling time is determined by simulation of an ideal behavioral model in the mixed-signal simulator Saber [Sab96] for a frequency step of 104 MHz and a settling accuracy of 100 Hz. In Fig. 5.5 (e), the settling simulation are plotted for three different bandwidths; For $\omega_c = 25$ kHz, 35 kHz and 45 kHz, the simulated settling time is 477μs, 293μs and 187μs respectively. Fig. 5.5 (f) shows a hyperbolic extrapolation of the simulated settling time over a larger bandwidth range. The settling can be normalized to the DCS-1800 spec (95MHz step and 180Hz accuracy) using the first-order formula (Eq. (2.33)), giving 278μs for the 35kHz loop filter. The settling performance at 35kHz is sufficient and can be enhanced by a higher basic charge pump current in measurements.

The second optimization step involves the choice of the charge pump current factor B and the fourth pole factor γ. The simulation results are plotted in Fig. 5.6 for $\omega_c = 35$kHz and $I_{qp} = 1.3\mu$A; Increasing the values for B and γ leads to better phase noise, but also to more

Loop Parameters		Loop Passives		Performance (at 600 kHz)	
ω_c	35 kHz	R_p	3.2 kΩ	$\mathcal{L}_{qp,z}$	-164.1 dBc/Hz
I_{qp}	1.3 μA	R_4	1.07 kΩ	$\mathcal{L}_{qp,p}$	-148.4 dBc/Hz
B	12	C_p	240 pF	$\mathcal{L}_{add}$	-139.8 dBc/Hz
α	4	C_4	710 pF	$\mathcal{L}_{int}$	-141.1 dBc/Hz
β	6	C_z	450 pF	$\mathcal{L}_{R_p}$	-137.8 dBc/Hz
γ	3	C_{tot}	1.4 nF	$\mathcal{L}_{R_4}$	-133.5 dBc/Hz
N	72			$\mathcal{L}_{vco}$	-127.5 dBc/Hz
PM	57°			$\mathcal{L}_{tot}$	-125.9 dBc/Hz
ζ	0.77			Settling time	293 μs

Table 5.3: *Summary of the loop properties and performance of the 4th-order, type-II PLL*

integrated capacitance. As a trade-off, the values $B = 12$ and $\gamma = 3$ are chosen (see the marker in Fig. 5.6).

A summary of the final PLL parameters and the results is given in Table 5.3. For the charge pump, an on-time fraction α_{qp} of 0.075 and a $V_{GS} - V_T$ of 0.3 V is chosen. The transconductance g_m of the filter adder and integrator are taken 2mS. Fig. 5.7 (a) shows the simulated open loop gain. The crossover frequency ω_c is indeed 35 kHz. The phase margin is 57° and the damping factor ζ is 0.77, which is slightly overdamped to avoid excessive ringing and slow settling. In Fig. 5.7 (b), the root-locus of the 4th-order, type-II loop is presented; The open loop poles and zeros are clearly visible: two poles at the origin, one low-frequency zero at 8.8kHz and two high-frequency poles at around 210kHz. For increasing loop gains, the one low- and one high-frequency pole come together and form a complex conjugate pair. For very high gains, the loop becomes unstable, since the conjugate pair moves into the right half-plane. The operating point of the presented PLL is marked by the pluses.

In Fig. 5.7 (c)-(f), the respective phase noise contributions of all loop components is plotted. (c) shows clearly that the VCO phase noise is high-passed to the output and determines the PLL for higher offset frequencies. However, the in-phase noise performance is also affected by the VCO; In the first place, the VCO exhibits much $1/f^3$ phase noise, which shows up in the PLL band and in the phase noise bump at intermediate offset frequencies. As discussed in Section 4.6.2, the VCO is not designed for low flicker noise upconversion, but for ultra low phase noise at 600 kHz and 3 MHz. As a result, the $V_{GS} - V_T$ of the g_m-transistors is taken rather high, which result in a higher flicker noise upconversion. The noise at low offset frequencies is mainly determined by the reference Xtal oscillator (see (d)). Fig. 5.7 (e) confirms that the passive loop components degrade the noise at 600kHz; Especially R_p is dominant. The active loop components are less critical at higher offset frequencies. However, the 1/f noise of the charge pumps deteriorates the in-band PLL phase noise and should therefore always be taken into account. The charge pump noise $I_{qp,z}$ normally has a flat transfer function within the loop bandwidth, but the 1/f noise increases the noise to critical levels, as shown in Fig. 5.7 (f).

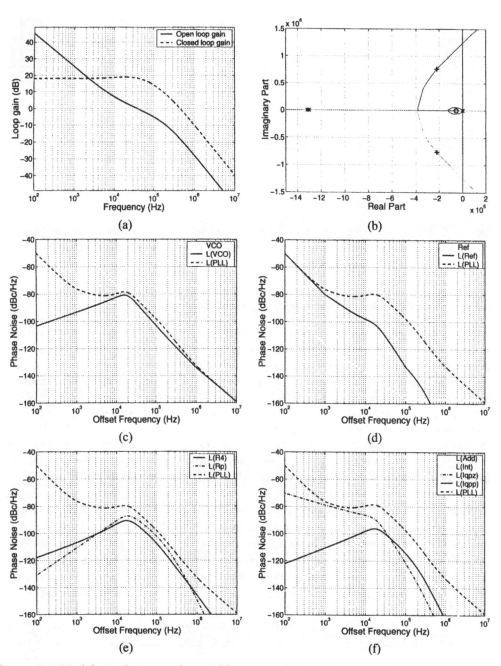

Figure 5.7: *Matlab simulation results: (a) The open and closed loop gain and (b) the root-locus of the 4th-order, type-II dual-path PLL. Phase noise contributions of (c) the VCO ,(d) the Xtal reference, (e) the passive loop components and (f) the active loop elements and charge pumps*

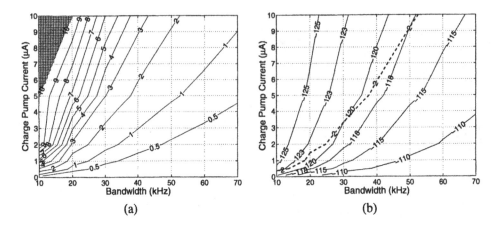

Figure 5.8: *Matlab simulation results for a type-II, 4th-order loop with the conventional filter of Fig. 5.2: (a) The integrated capacitance (nF), (b) the phase noise at 600 kHz (dBc/Hz) for N=65 compared with the 2nF capacitance line (dashed).*

5.3.4 Conventional Versus Dual-Path Topologies

In Section 5.2.2, some numerical examples were given for a 4th-order, type-II PLL with the conventional active loop filter of Fig. 5.2. The results showed that to keep the phase noise within acceptable levels, the necessary capacitance rises up to 6nF! To make a comparison between the conventional topology and the dual-path loop filter topology, simulations similar to those of the previous section are performed for the bandwidth and the charge pump current. To make an honest comparison, a similar equalization circuit is incorporated in the simulation of the conventional loop and the simulations are performed at the worst case division modulus, i.e. $N = 65$. The factors α, β and γ are equal to that of the dual-path topology for stability reasons. The g_m of the integrator opamp is 4mS, to get the same power consumption as in the dual-path case. The optimization goals are the same as before: 2nF, -123 dBc/Hz at 600 kHz, 2° and 288μs.

The results of the simulation are shown in Fig. 5.8. In Fig. 5.8 (a), the integrated capacitance is plotted in nF and the shaded area is again forbidden territory. The capacitance in a fixed simulation point is smaller for the conventional loop filter topology. In Fig. 5.8 (b), the phase noise at 600 kHz is presented together with the capacitance spec line at 2nF. The rms phase error and settling time of the conventional topology is similar to that of the dual-path topology, such that the bandwidth should be at least larger than 25 kHz. As can be seen, it is impossible to satisfy the optimization goals for any combination of bandwidths and charge pump currents. To make a loop filter with a bandwidth of 35 kHz for settling and rms phase error, the capacitance needs to be at least 5.8nF to satisfy the noise specification. With a 2nF capacitance, a phase noise spec of maximally -120 dBc/Hz can be obtained which is unacceptable, keeping in mind the effort put in optimizing the VCO for the lowest phase noise. In short, for the same noise spec, the integrated capacitance of a conventional filter is more than a factor 4 higher than for the dual-path loop filter implementation, proving the effectiveness of the dual-path principle.

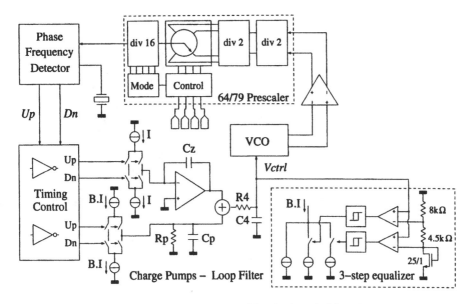

Figure 5.9: *The circuit implementation of the dual-path 4th-order, type-II PLL.*

5.4 The PLL Building Block Circuits

In Fig. 5.9, the full PLL circuit is depicted. The VCO circuit is not shown, since it is elaborately discussed in Section 4.6.2. The prescaler block diagram is shown, but implementation details can be found in Section 3.5. The phase-frequency detector circuit together with the charge pumps and the timing control to maximize the suppression of spurious tones is elaborated in the next chapter in Section 6.8. This order of presentation has been chosen since the design of the PFD and the charge pumps is critical for the correct operation of the PLL in $\Delta\Sigma$ fractional-N mode. In what follows, the implementation of the loop filter and of the 3-step equalizer is discussed. The buffer between the VCO and the prescaler is implemented as two simple common-source, single-transistor amplifiers and prevents prescaler kick-back noise from entering the LC-tank of the VCO . All building blocks are designed to operate at a power supply voltage from 1.8V to 2V. The low power supply voltage is chosen to prove the feasibility of low-voltage RF CMOS for compatibility with the digital part in view of single-chip transceiver integration.

5.4.1 The 3-step Equalizer Circuit

As shown in Section 4.6.2, the tuning curve of the VCO is rather non-linear, resulting in a variation of the derivative, i.e. the VCO gain K_{vco}; The K_{vco} ranges from 300-120 MHz/V. Since the VCO gain influences the open-loop gain and thus the loop bandwidth and the phase margin (Eq. (5.14)-Eq. (5.16)), the gain variation can result in loop instability. A way has to be found to equalize the overall loop gain over the frequency range.

The most straightforward method to perform a continuous equalization is to pre-process the

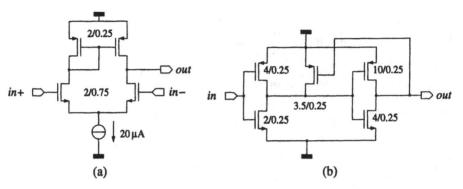

Figure 5.10: *The circuit schematics of (a) the amplifier and (b) the Schmitt-trigger in the 3-step equalizer.*

V_{ctrl} (V)	K_{vco} (MHz/V)	I_{qp} (μA)
< 0.6V	300	1.3
< 1.2V	200	2
< 2V	120	3.2

Table 5.4: *The three equalization steps and the corresponding charge pump currents. The relationship between the currents is approximately $\sqrt{2.5}$.*

control voltage V_{ctrl} at the filter output, such that it compensates the non-linear tuning characteristic of the VCO. The non-linear I/V characteristic of MOS transistor can be used for this purpose. However, both non-linear curves must be inversely matched. Additionally, the gain range must be maintained over threshold, transconductance and resistance variations. Last but not least, the noise of this circuit has the same transfer function as the VCO and appears as out-of-band phase noise. Several tens of mA are necessary to lower the circuit noise to negligible levels.

A more applicable solution is to not equalize the VCO gain, but to equalize the open loop gain of the PLL. To keep the open loop gain (Eq. (5.14)) constant, it is sufficient to equalize the product $I_{qp} \cdot K_{vco}$. In this design, the equalization is implemented in a 3-step piece-wise way, by adding and removing current sources from the output current $B.I$ as shown in Fig. 5.9 in the lower right corner. Since the VCO gain ranges over a factor 2.5, the charge pump current must range from 1.3μA to 3.2μA. For decreasing K_{vco}, the charge pump current is increased in logarithmic steps, meaning that every step the current is increased by a factor $\sqrt{2.5}$. The equalization steps and the corresponding charge pump currents are summarized in Table 5.4. The high-level equalization circuit is shown in Fig. 5.9. The tuning voltage to the VCO V_{ctrl} is compared to 2 reference levels by two very simple amplifiers (see Fig. 5.10 (a)). Since the specs for the amplifier are not severe, only 20μA is needed for their operation. The outputs of the amplifiers is used to switch the different current sources. Problems arise when V_{ctrl} is close to one of the reference voltages; Adding too much current by switching a current source on, can esult in loop instability. Moreover, the PLL can get stuck in a regime were the current sources re constantly switched on and off. To avoid this regime, Schmitt-triggers are inserted after the ·pamps to introduce hysteresis in the switching process. The circuit of Fig. 5.10 (b) provides round 20 mV hysteresis.

The output signals of the Schmitt-triggers control the switches of the current bank. The left ·urrent source delivers the $B \cdot 1.3\mu A = 15.6\mu A$, the second one adds $8.4\mu A$ and the last one .4.4μA. All gate lengths of the current source transistors are $20\mu m$ for current matching, high ·utput resistance and low 1/f noise. The switches have minimum length for small resistance. The ·utput current of the equalization circuit is directly fed to the charge pumps.

The equalization not only improves the stability of the loop over the frequency range but also he phase noise performance. First, the noise from the equalization circuit itself is not important; n the locked state, both the up and down current are active for α_{qp} of a reference period, but he noise contribution of the equalization circuit is equal for both current sources and therefore loesn't show up at the charge pump output. Secondly, lowering the K_{vco} results in lower phase noise. If on top of that the charge pump current is increased inversely with K_{vco}, an additional mprovement is realized. If for example the VCO gain is halved, I_{qp} is doubled. The noise rom the charge pump currents increases by 3dB but is not critical for the output phase noise. However, it can be proven from Eq. (5.23) and Eq. (5.22) that the phase noise of R_4 and R_p at higher offset frequencies is proportional to K_{vco}/I_{qp}, such that the noise contribution of R_4 and R_p is reduced by 6dB at the high-frequency end of the PLL range. As a result, the overall output phase noise is improved by more than 1 dB due to the equalization. This improvement does not occur in the lower frequency range, emphasizing the worse phase noise at $N = 65$ even more compared to the $N = 72$ case, as observed in Fig. 5.5 (d).

5.4.2 The Loop Filter Circuit

As explained in Section 5.3.1, the dual-path loop filter requires that two filter paths, i.e. one active integration and one passive pole, are added. The high-level dual-path loop filter implementation is shown in Fig. 5.4 and in Fig. 5.9. It must be noted that the diode capacitance of the VCO is in parallel with C_4 and changes depending on the operating point of the PLL. Therefore, the loop gain, phase margin and cross-over frequency change. However, the diode capacitance varies from 3.1pF to 2.5pF in the DCS-1800 band, resulting in a variation of less than 0.1% on the 710pF C_4 capacitance.

The loop filter opamp circuit to implement the active integration is presented in Fig. 5.11 (a). The active implementation was necessary to keep the charge pump output at V_{ref} to avoid mismatch in the up- and down-currents and reduce spurious charges. A PMOS differential pair and PMOS current source is chosen for low 1/f noise. All transistors are made large for the same reason. The circuit consumes $270\mu A$ and is designed to have a $g_{m,Int}$ of 2mS with a gain of around 85. Since the supply voltage is sub-2V, the loop filter must be able to deliver an as high as possible voltage range to tune the VCO. Therefore, all PMOS transistor have a $V_{GS} - V_T$ of only 0.125V.

The loop filter adder circuit is shown in Fig. 5.11 (b). The signal from the passive filter part $-B.V_p$ is converted to a current by transistor Mp. The signal of the active path V_z is converted

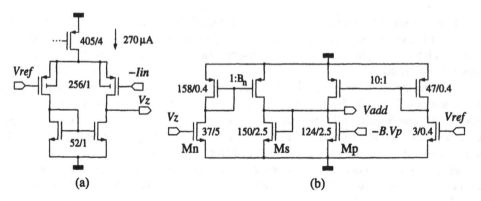

Figure 5.11: *The circuit schematics of (a) the amplifier and (b) the adder in the dual-path loop filter of Fig. 5.4.*

to a current signal by Mn and mirrored to the output with a factor B_n. Both currents are summed and converted to the output voltage V_{add} by the g_m of the NMOS diode Ms. Since the both current paths have opposite signs, the voltage are effectively summed as required by Eq. (5.13). A reference current branch is added to the circuit to set the proper DC operating point of the filter output. To satisfy the upper part of the tuning range the $V_{GS} - V_T$ of the PMOS transistors is again chosen minimal, i.e. 0.125V. This means that the output voltage can go up to 1.875V. The lowest possible output voltage is determined by Ms. Since the gate and drain of the diode transistor Ms are connected, the output voltage is determined by its V_{GS}. Therefore, the $V_{GS} - V_T$ is again 0.125V. With a V_T for the NMOS of 0.485V, resulting in a minimum output voltage of 0.61V. This means that the calculated output voltage range is from 1.875V to 0.61V, which is sufficient to synthesize the DCS-1800 frequency band. Simulations predict an even higher voltage range of 1.88V to 0.25V.

The transfer function of the adder circuit is as follows:

$$V_{add} = \frac{1}{g_{m,s}} \cdot \left(B_n \cdot g_{m,n} \cdot V_z + g_{m,p} \cdot B.V_p \right) \tag{5.25}$$

To make the addition, the transfer functions of both inputs to the output should be one. This translates to $g_{m,p} = g_{m,s}$ and $g_{m,n} = g_{m,s}/B_n$. In this design case, transconductance of the summing transistor Ms, $g_{m,s}$ is chosen to be 3mS for low noise and B_n is 6 for low power. The overall current consumption of the circuit is 2.4mA, mainly to keep its phase noise contribution low.

5.5 Experimental Results

Based on the results of the optimization process in Section 5.3.3, a phase-locked loop frequency synthesizer is integrated in a standard 0.25μm CMOS process with only two metal layers. The

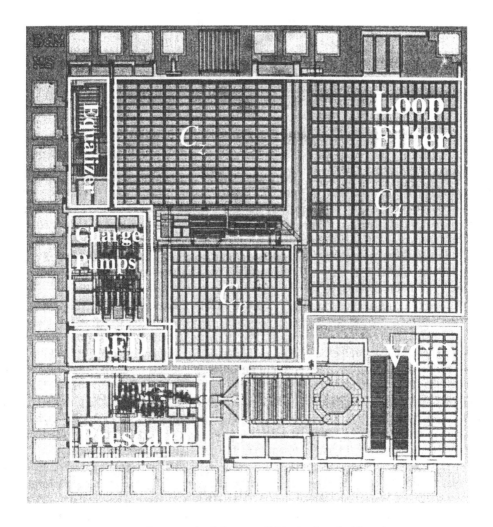

Figure 5.12: *IC microphotograph of the monolithic dual-path 4th-order, type-II PLL.*

	Power (mW)
PFD	0.2
Charge pumps	1
Loop filter	8
VCO	46
Prescaler	10
Equalizer	1
VCO buffers	2
Output buffers	2
Total	**70.2**

Table 5.5: *The power distribution of the PLL building blocks for 2V.*

IC microphotograph is presented in Fig. 5.12; The IC features a fully integrated 35 kHz dual-path loop filter, a fully integrated LC-tank VCO, a high-speed 64/79 prescaler, a zero-dead-zone phase-frequency detector, a 3-step equalizer and dual charge pumps. The VCO and prescaler re-use the layout of the active part of the implementations presented in Section 4.6.2 and Section 3.5. The lion's share of the IC is occupied by the three loop filter capacitors: C_z (450pF), C_p (240pF) and C_4 (710pF). They are realized using the metal/poly structures available in the CMOS process with a nominal capacitance value of 1.3 fF/μm ^{2}. The total die size of the monolithic PLL is 2×2mm^2, including bonding pads and bypass capacitors.

All measurements are performed with a power supply voltage of only 2V. The PLL is even fully functional at 1.8V, except for the synthesizable frequency range, which is reduced, since the filter voltage output range is compressed. The reference frequency of 26 MHz is delivered by an off-chip temperature compensated Xtal oscillator. The power consumption of the total IC is 70mW at 2V. The distribution of the power of the different building blocks is listed in Table 5.5. Note that the fully integrated, low-phase-noise VCO is responsible for more than 66% of the total power consumption. The output buffers are placed at the VCO output, the prescaler output and the V_{ctrl} for measurement purposes. The digital part of the IC, including the high-speed prescaler, consumes only 16.2mW. At 1.8V, the power consumption is still 57mW, since the VCO needs approximately the same current to maintain its low-phase-noise operation.

The phase noise performance of the PLL is measured with a dedicated phase noise measurement system with the delay line method. Due to a pole in the delay line the measurements lose accuracy for offset frequencies higher than 5 MHz. The phase noise measured at 1.82 GHz is plotted in Fig. 5.13 (a) and compared with the simulated phase noise around 1.82 GHz ($N = 70$). The simulations are in excellent agreement with the measurements, except for the in-band noise which is a little higher than simulated. The first reason for the discrepancy could be 1/f noise that is higher than expected, since the 1/f noise model parameters are not accurate and the 1/f^3 noise from building blocks other than the VCO is not taken into account. A second reason is the mismatch between the implemented and actual absolute values of the passive loop filter components. The variance on the absolute values is up to 5%, which in a worst-case situation leads to a 10% reduction or increase in PLL bandwidth. At 600 kHz offset the phase noise is as low

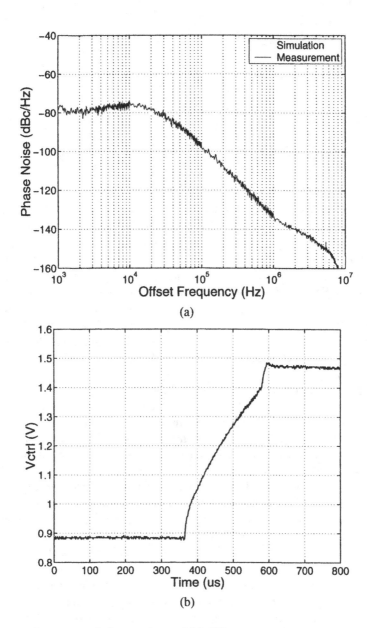

(a)

(b)

Figure 5.13: *(a) The measured phase noise at 1.82 GHz compared to the simulated phase noise (gray). (b) The measured loop dynamics of the PLL for a 104 MHz step.*

	DCS-1800	Measured
Phase Noise at 600 kHz (dBc/Hz)	-116	-125.8
Phase Noise at 3 MHz (dBc/Hz)	-133	-143.7
$\Delta\Phi_{rms}$	$2°$	$1.8°$
Settling Time (μs)	865/288	215
Reference Spurs (dBc)	-80	< -84
Power Consumption (mW)	–	70
Power Supply (V)	–	2
Area (mm ^{2})	–	2×2
Frequency Resolution (Hz)	200 k	26 M

Table 5.6: *Summary of measured specifications compared to the DCS-1800 specifications*

as -125.8 dBc/Hz and -143.7 dBc/Hz at 3 MHz. The phase noise at 3 MHz is better than the VCO noise taken from simulations in Section 5.3.3; The bias current of the VCO implemented in the PLL is higher, resulting in higher power consumption but also better out-of-band phase noise of the PLL, due to the higher oscillation amplitude. The phase noise at the lower end of the frequency range for $N = 65$ is worse than simulated due to the higher VCO noise (around -124dBc/Hz at 600kHz) and $1/f^3$ noise, but still within the DCS-1800 spec (see Table 5.6).

The spurious suppression at the PLL output is measured to be less than -84 dBc, which is the noise floor of the spectrum analyzer. In other words, no reference spurious tones could be distinguished. The excellent spurious suppression is owed to the careful charge pump design with timing control (see Section 6.8), abundant use of bypass capacitors and separation of the different power supplies.

The loop dynamics of the PLL are measured at the filter output by buffering with a source-follower. Therefore the measured settling characteristic is level-shifted compared to the original one. The result is plotted in Fig. 5.13 (b). The measured settling time for a 104 MHz step (a change of 4 division moduli) is 226μs, which is faster than simulated since in the measurements a higher charge pump current of 2μA is employed to further decrease the settling time and the in-band phase noise. The measured settling time comes down to 215μs for a 95 MHz step and an accuracy of 180 Hz for DCS-1800. The overall settling characteristic corresponds to the simulated one in Fig. 5.5; The slope of the measured settling is less linear than the simulated one due to the equalizer, which was not incorporated in the simulations. The output frequency however changes linearly for a modulus step. Other differences between measurements and simulations are attributed to variations in the loop passives and to the lack of non-idealities in the simulations.

To conclude the experimental results, the measured specifications are summarized in Table 5.6 and compared to the DCS-1800 class I/II system specifications derived in Section 2.6. The optimization process, presented in this chapter, has been targeted towards low phase noise, to show the feasibility of low phase noise in a fully integrated solution. As a result, the measured phase noise exceeds the DCS-1800 with 10 dB, which means that the noise power of the PLL is ten times lower than required. The rms phase error is with $1.8°$ sufficiently lower than the required spec, but higher than the simulated $1.57°$ at $N = 70$ for the reasons stated above. The excellent phase noise performance comes at the expense of an increased power consumption, in spite of the low supply voltage of 2V; This is mainly due to the VCO, whose phase noise is not voltage- but current-limited because of the single PMOS topology. The area of the IC is within reasonable measures, due to the careful optimization of the noise-capacitance trade-off. The settling time and reference spur suppression are also well within spec. The frequency resolution is now determined by the reference frequency, i.e. 26 MHz, since the PLL is operated in integer mode. In the next chapter, the PLL is used as a $\Delta\Sigma$ fractional-N synthesizer for DCS-1800 resulting in a frequency resolution of 400 Hz.

5.6 Conclusion

In this chapter, a fully operational prototype of a monolithic phase-locked loop frequency synthesizer for the DCS-1800 class I/II communications standard, integrated in a standard $0.25\mu m$ CMOS process has been discussed. The IC features a fully integrated 35 kHz dual-path loop filter, a fully integrated LC-tank VCO, a high-speed 64/79 prescaler, a zero-dead-zone phase-frequency detector, a 3-step equalizer and dual charge pumps. Moreover, the VCO is capable of driving a polyphase filter in a power efficient manner. These realizations pave the way to the integration of a complete phase-locked loop in CMOS, without trimming, tuning and external components. The PLL operates from a power supply voltage of 2V to prove the feasibility of low-voltage RF CMOS for compatibility with the digital part in view of single-chip transceiver integration.

Since the high-speed 16-modulus prescaler and a high-quality, low-phase-noise VCO are considered off-the-shelf available, the focus of this chapter is on the feasibility of fully integrated loop filters. The loop filter design emphasizes the minimization of the integrated capacitance while maintaining and exceeding the targeted phase noise performance. As a starting point, the conventional 3rd-order, passive charge pump PLL was evaluated and rejected for its lousy noise performance. Increasing the order and reverting to an active implementation, proved to significantly improve the spectral purity, but at the same time blow up the necessary capacitance... as could be expected from the capacitance-noise law of nature. To relax the inherent trade-off, a dual-path loop filter topology has been developed; By combining two filter paths with a relative current multiplication, the area-expensive capacitance that realizes the zero does not need to be integrated as such. A design strategy has been elaborated for the dual-path loop filter that enables to demarcate and optimize the available design space within noise, settling and area constraints over the full frequency range. With the design strategy, a 35 kHz dual-path loop filter has been implemented that effectively reduces the necessary capacitance for the same noise with more than a factor 4 as compared to the conventional topologies!

The loop filter and charge pump parameters have been set by the optimization strategy. The total integrated capacitance is 1.4nF, which is still significant due to the high VCO gain at the lower frequency end. The high K_{vco} amplifies the noise of the loop filter resistors and degrades the noise at 600 kHz offset. Moreover, the variation of the K_{vco} over the frequency range is quite high, necessitating a compensating equalization circuit, to maintain the PLL stability over the full frequency range. The PFD and dual charge pump circuits have been designed for maximum spurious suppression (see Section 6.8). The most critical circuit in terms of low-voltage operation is the loop filter adder; A current adding topology has been chosen to maximize the tuning voltage range for the VCO.

The monolithic 4th-order, type-II PLL was realized on a single 2×2 mm^2 die in a standard, 2 metal layer 0.25μm CMOS process, without external components and without tuning, trimming or post-processing. The IC consumes 70mW from a 2V power supply. The measurements show full compliance with the stringent DCS-1800 class I/II specifications. The measured phase noise even exceeds the DCS-1800 specs with over 10 dB. The measured PLL phase noise performance at critical offset frequencies exceeds that of most published fully integrated VCOs! The measured dynamic behavior, the rms phase error and phase noise agree well with the simulated values extracted from the developed design strategy. The in-band noise is slightly higher than expected which is attributed to 1/f noise and 1/f^3 noise, which is not incorporated in the simulations, and process variations, although measurements of different samples show little variation in the PLL performance.

An unambiguous comparison of the presented PLL implementation and other work is difficult, since the PLL performance is determined by many different specifications and most synthesizer implementations in open literature do not achieve the same, high degree of integration. In most cases, only the digital PLL part is integrated on-chip. The few monolithic synthesizers in open literature are mostly aimed for low-end applications, such as Bluetooth [Leen02], and present no competition for the presented monolithic synthesizer. Only the synthesizer implementation of [Cran98] aims for DCS-1800 cellular specifications. The phase noise of this 0.4μm CMOS PLL is more than 3 times (5dB) worse than the presented synthesizer with only 30% less power and comparable area.

In the presented implementation, monolithic integration combined with an excellent performance was mainly possible through the high reference frequency, i.e. 26 MHz. A high bandwidth of 35 kHz could be implemented which is beneficial for in-band noise as well as settling and integratability. However, the frequency resolution of the PLL is limited to 26 MHz. In the next chapter, the presented PLL is extended for use as a $\Delta\Sigma$ fractional-N synthesizer, resulting in a frequency resolution of 400 Hz.

Chapter 6

A 1.8 GHz CMOS $\Delta\Sigma$ Fractional-N Frequency Synthesizer

6.1 Introduction

The architectural simplicity of integer-N PLL frequency synthesizers has made them a popular choice for all kinds of telecommunication systems. However, the integer-N architecture has a major drawback: the frequency resolution, i.e. the channel spacing is equal to the reference frequency, meaning that only integer multiples of the reference frequency can be synthesized. Stability requirements limit the loop bandwidth to about one tenth of the reference frequency. To ensure sufficient suppression of spurious signals at the reference frequency, the loop bandwidth needs to be even smaller. As a result, the dynamic behavior of the PLL is seriously degraded. In addition, a high division modulus N is necessary. Since the noise contributions of almost all PLL building blocks, except the VCO, are multiplied by N, the in-band noise of the PLL –and thus the rms phase error– becomes unacceptable. In short, the design of integer-N PLL frequency synthesizers poses a severe trade-off between frequency resolution, spectral purity – spurious as well as noise – and PLL dynamic behavior.

During the years, several attempts have been undertaken by design engineers to decouple the different PLL specifications. One of the first attempts to decouple fine frequency resolution and dynamic behavior is the multi-loop frequency architecture [Raz97b]. One loop generates a fixed, high-frequency carrier to which a combination of low-frequency signals is added by SSB mixing (see Chapter 2). However, the multi-loop architecture requires accurate SSB mixing to keep sidebands lower than -60 to -70 dBc. The mixer also has a frequency dependent delay, which could introduce loop stability problems.

One of the most promising techniques to supply more degrees of freedom in synthesizer design is fractional-N synthesis [King75, Mill90, Ril93]. In fractional-N frequency synthesizers, *fractional* multiples of the reference frequency can be synthesized, allowing a higher reference frequency for a given frequency resolution, which means that the loop bandwidth can be increased, without deteriorating the spectral purity. Therefore, the PLL dynamics are accelerated and the total amount of required capacitance in the loop filter can be decreased, such that sin-

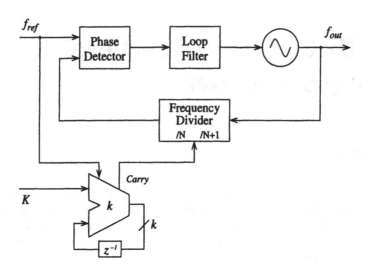

Figure 6.1: *The Fractional-N Principle.*

gle chip integration of the frequency synthesizer becomes feasible. The fractional-N synthesizer employs the same, straightforward architecture as the integer-N synthesizer, with additional digital logic to properly control the division moduli. Frequency resolution is traded against digital complexity. Since CMOS technology is the natural biotope of digital circuitry, the fractional-N synthesizer is an excellent option for full CMOS frequency synthesizer integration.

6.2 The Fractional-N Principle

The basic idea behind fractional-N synthesis is division by *fractional* ratios, instead of only integer ratios [King75]. To accomplish fractional division, the same frequency divider as in an integer-N frequency synthesizer is employed, but the division is controlled differently. In Fig. 6.1 the division modulus of the frequency divider is steered by the carry output of a simple digital accumulator of k-bit width. To realize a fractional division ratio $N+n$, with $n \in \mathbb{R}[0, 1]$, a digital input $K = n \cdot 2^k$ is applied to the accumulator. A carry output is produced every K cycles of the reference frequency f_{ref}, which is also the sampling frequency of the digital accumulator. This means that the frequency divider divides $2^k - K$ times by N and K times by $N + 1$, resulting in a division ratio N_{frac}, given by Eq. (6.1).

$$N_{frac} = \frac{(2^k - K) \cdot N + K \cdot (N + 1)}{2^k}$$

$$= N + \frac{K}{2^k} = N + n \tag{6.1}$$

Eq. (6.1) states that for a given reference frequency, it is possible to make the frequency resolution arbitrarily fine, by choosing the width of the accumulator sufficiently large. For example, in a

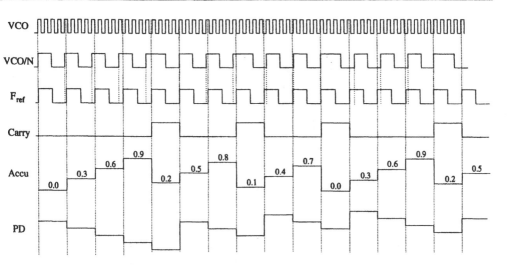

Figure 6.2: *The chronograph of the VCO and frequency divider output, the overflow bit, the accumulator output and the phase detector output for N = 4 and n = 0.3.*

DCS-1800 telecommunication system the channel spacing of 200 kHz can be synthesized using a f_{ref} of 26 MHz, by realizing an accumulator width k of more than 7 bits.

Since the fractional division is applied to a *phase*-locked loop, it is important to investigate what happens to the phase in the loop, to fully understand the effects on the overall loop performance. Each cycle of the reference clock, the accumulator accumulates the input K, until the accumulator output overflows. At overflow, the carry output is set high and 2^k is subtracted from the accumulator output. The resulting accumulator output is a sawtooth waveform with a frequency depending on n, as shown in Fig. 6.2 for $n = 0.3$. The carry output of the accumulator is used to control the division ratio of the divider. The carry output has a mean value of n, but the instantaneous value is merely a "prediction" of the mean value, since the output can only be 0 or 1. Therefore, the frequency divider is dividing only by integer ratios, N or $N + 1$, while the output of the PLL is a fractional multiple of f_{ref}. When the divider is dividing by N, the divider output leads the reference frequency, giving rise to an increasing phase difference between both phase detector inputs. The phase detector output current starts to accumulate. When the accumulator overflows, the divider modulus is changed from N to $N + 1$, meaning that one output period of the VCO is swallowed by the frequency divider (the shaded pulses in Fig. 6.2). This causes a phase shift at the input of the phase detector that eliminates the previously accumulated phase error. Fig. 6.2 shows the chronograph of the open-loop output of the phase detector versus the VCO output and f_{ref} for $N = 4$ and $n = 0.3$.

The accumulator, as shown in Fig. 6.1, acts as a *phase* accumulator; When dividing by N, the divider output period T_{div} equals $N \times T_{out}$, while the reference period T_{ref} always equals $N_{frac} \times T_{out}$. Since T_{ref} is larger than T_{div}, the divider output leads the reference clock with $\Delta t = (N_{frac} - N) \times T_{out}$ every reference cycle. The corresponding phase error increase is than

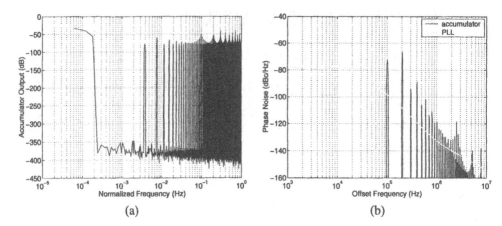

Figure 6.3: *The output spectrum of an 8-bit accumulator (a) and the effect on the phase noise spectrum of the PLL of Chapter 5 (b) for $N = 64$, $n = 0.3$ and $f_{ref} = 26$ MHz with ideal PLL building blocks.*

$\Delta\theta = -n \cdot 2\pi/N_{frac}$. By convention the phase error is taken negative when the divider output leads the reference clock. When the carry out is 1, the divider period equals $(N + 1) \times T_{out}$, meaning that Δt becomes $(N_{frac} - N - 1) \times T_{out}$. The phase error decreases by $\Delta\theta = -(n - 1) \cdot 2\pi/N_{frac}$. Regardless of the division ratio, the phase error changes by $\Delta\theta = -(n - co) \cdot 2\pi/N_{frac}$, with co the carry output of the accumulator. Therefore, the instantaneous phase error θ_{err} is given by:

$$\theta_{err} = \frac{2\pi}{N_{frac}} \sum_i (co[i] - n) \tag{6.2}$$

Eq. (6.2) shows that the phase error is a time-integrated scaled version of the output of the digital accumulator. In fact, the output of the accumulator is an inverted version of the phase error, advanced by one period of f_{ref}. Therefore, the output of the accumulator register is a measure for the phase error. The register is also called the *phase register*. Note that the fractional-N PLL is never really locked, since the phase error is never zero over more than one period of f_{ref}. The term fractional-N PLL is somehow misleading, because the loop never provides a fractional division ratio, but rather changes its division ratio over time to provide a mean division by a fraction.

The pulse swallowing action, performed by the frequency divider/accumulator circuit, is a periodic action, i.e. every $2^k - K$ cycles of f_{ref} a VCO pulse is swallowed. Therefore, the sawtooth-shaped phase error is also periodic with a period depending on K. The resulting AC component, superimposed on the wanted DC output of the phase detector, modulates the VCO frequency and large spurious components appear in the PLL output spectrum at multiples of $n \times f_{ref}$, even after filtering by the loop filter. As an example, a simulation is performed where an 8-bit digital accumulator controls the least significant division modulus of the PLL, discussed in Chapter 5. The frequency spectrum of the accumulator output, normalized to represent the actual phase error within the loop is shown in Fig. 6.3(a). The high spurious content seriously corrupts

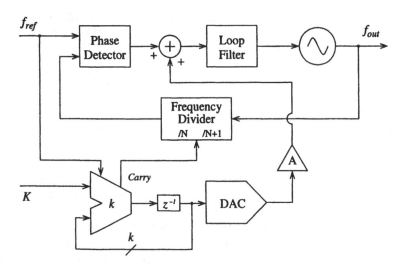

Figure 6.4: *The analog phase interpolator: the output of the phase register is converted to an analog signal by the DAC and added to the phase detector output to cancel the sawtooth waveform on the phase error. A scaling block A is inserted to increase the cancellation accuracy.*

the PLL output spectrum (see Fig. 6.3(b)), even for an ideal PLL. The problem gets worse when going to smaller values of n. If for example $n = 0.001$, the spurs, larger than -30 dBc, appear at multiples of 26 kHz.

6.3 Conventional Fractional Compensation Methods

6.3.1 The Analog Phase Interpolator

At first glance, the fractional-N frequency synthesizer seems to have unlimited advantages. It manages to decouple the frequency resolution and the bandwidth-settling requirements of an integer-N PLL, without much added complexity. Unfortunately in reality, the fractional-N synthesizer design is still a compromise between resolution and bandwidth, be it less harsh than for its integer-N counterpart. To suppress the large spurious energy at multiples of $n \times f_{ref}$, the loop bandwidth has to be smaller than would normally be required for the same f_{ref}. A typical value for the loop bandwidth of fractional-N synthesizer is $f_{ref}/100$ versus $f_{ref}/10$ in integer-N synthesizers [Gold95]. The trade-off can not be resolved by increasing f_{ref}, due to the limited operating frequency of the phase detector circuitry.

To achieve the spectral purity required by modern high-quality communication systems, techniques have to be found to cancel or remove the AC phase error, that modulates the VCO, such that only the wanted DC phase error remains. The oldest and most straightforward technique is the *analog phase interpolator* [Cox76, Rhod83]. As proven above, the output of the phase register of the digital accumulator is a measure for the phase error. The phase register is a kind of "bookkeeping system" that tracks the phase advancement of the VCO for every reference cycle. Due to the loop integration the actual phase detector output becomes an analog sawtooth waveform. Using a DAC (Digital-to-Analog Converter), the staircase output of the phase register can be converted to an analog sawtooth current, scaled by A (Fig. 6.4) to match the phase error. By summing the phase detector output and the DAC output on the loop filter capacitors, the AC component is in the ideal case removed from the phase error (see Fig. 6.4). The VCO output is now only driven by the wanted DC phase error, cancelling all fractional spurs in the output spectrum. This method has several disadvantages. First of all, the cancellation of the spurs is usually performed in the current domain and the compensation current from the DAC can seriously deteriorate the in-band phase noise of the synthesizer. Secondly, the requirements posed on the DAC are severe. First of all, the maximum amplitude of the phase error is dependent on f_{ref} (lower amplitude for higher frequencies), such that the DAC output amplitude must be variable, while its input is constant. Secondly, the sampling frequency of the DAC can be rather high. To have fast settling (large bandwidth), the f_{ref} can easily be some tens of MHz. The sampling frequency of the DAC must be at least $2 \times f_{ref}$, for a full Nyquist DAC, to be able to compensate spurs up to the reference frequency. To have a spurious suppression of at least -70 dB, required in most telecommunication systems, the accuracy of the DAC must be around 8 bits, for the PLL of Chapter 5 and 200 kHz frequency resolution. Integrating such a DAC in e.g. a 0.25 μm CMOS technology, 0.5 mm^2 of area is necessary and the power consumption is in the order of 10 mW. Thirdly, due to matching issues, the accuracy of the cancellation is limited, requiring external adjustment and calibration. Therefore, the analog phase interpolator is not a viable solution for full integration of frequency synthesizers for wireless communications systems in CMOS technology.

6.3.2   The Fractional Divider

Another technique that performs "fractional compensation", i.e. compensation of the fractional spurs, is the fractional divider based fractional-N synthesizer [Fran93]. The goal of the technique is to let the divider divide actually by fractional ratios, such that the reference clock and the divider output are in-phase and the loop is actually locked. The fractional divider requires the same hardware as the plain pulse swallowing fractional-N synthesizer (Fig. 6.1), but a DLL (Delay Locked Loop) or a similar circuit, that can alter phase in a programmable way, is inserted between the divider and the phase detector. The frequency divider division ratio is again controlled by the carry output of the accumulator. The phase error introduced by the coarse "prediction" of the phase is sensed by the DLL and compensated by changing the delay according to the phase error. Therefore, the number of delay units in the DLL must be equal to the minimal resolution of the accumulator, i.e. $1/2^k$. This technique requires again an excess of area and power consumption and still introduces phase error due to non-ideal compensation by the DLL. Also the in-band synthesizer noise can be seriously degraded, because the delay stages in the DLL are prone to high phase noise (see Section 2.4.3.3).

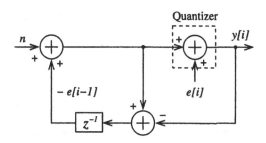

Figure 6.5: *The equivalent model of an accumulator.*

6.4 ΔΣ Modulation in Fractional-N Synthesis

6.4.1 Introduction

ΔΣ modulators in fractional-N synthesis were first introduced and analyzed by Miller et al. (HP) [Mill90, Mill91] and further refined by Riley et al [Ril93]. Since the ΔΣ modulator is a pure digital block in these systems, all analysis in this book makes a clear distinction between the digital domain and the analog domain in the PLL. The loop variable is phase in both domains. The analog part consists of the charge pump amplification, the loop filter and the VCO. All analysis of the analog part is performed in continuous time and in the frequency domain ($s = j\omega$). The digital part contains the division control logic (a ΔΣ modulator in this case), the multi-modulus prescaler and the PD or PFD. The PFD is modeled as a block that can output a current which is an arbitrary function of the phase difference, enabling non-linear analysis. The digital analysis is performed in the z-domain ($z = e^{j\omega T_{ref}}$), e.g. E(z), and using discrete-time behavior, e.g. e[i] the i-th sample of the e(t), to incorporate any non-linearities. The digital part of the PLL is considered a sampled system with sampling frequency f_{ref}.

6.4.2 The Accumulator as Noise-Shaping Quantizer

By investigating the accumulator in Fig. 6.1 more thoroughly, some interesting properties arise. Each cycle of the reference frequency f_{ref}, the accumulated value in the accumulator is added to the input revealing a number between 0 and 2^{k+1}. If this number is larger than 2^k, the carry is set high and 2^k is subtracted from the accumulated value. In Fig. 6.5 the corresponding equivalent model for the accumulator is drawn, normalized towards the number of bits. In fact, the accumulator carry output is a coarse, discrete-time "prediction" of its input K, normalized towards number of bits, n. So, by doing this coarse prediction, the accumulator generates a quantization error, which is the staircase waveform discussed earlier. If the comparator is further modeled as an addition of a quantization error e[i] between 0 and 2^k, the discrete-time behavior, as well as the transfer function of the accumulator can be described.

$$y[i] = n + (e[i] - e[i-1]) \tag{6.3}$$

$$Y(z) = n + (1 - z^{-1}) \cdot E(z) \tag{6.4}$$

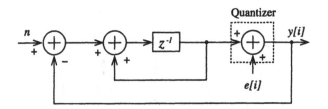

Figure 6.6: *The equivalent model of a first-order $\Delta\Sigma$ modulator.*

As can be seen from the equations, the carry output of the accumulator is indeed the input, with the addition of a *shaped* version of the quantization noise. The quantization noise is high passed, i.e. most error energy is situated around $f_{ref}/2$. This structure is in fact noise-shaping quantization, i.e. the well known $\Delta\Sigma$ technique [Nors97]. When comparing the equivalent model of a first-order $\Delta\Sigma$ modulator (Fig. 6.6) and the equivalent model of the accumulator (Fig. 6.5), both models can be shown to be algebraically equivalent. If the comparator in Fig. 6.6 is modeled as an addition of a quantization error $e[i]$, the transfer function of the accumulator can be derived as in Eq. (6.5).

$$Y(z) = z^{-1} \cdot n + (1 - z^{-1}) \cdot E(z) \qquad (6.5)$$

The output is a delayed version of the input with shaped quantization noise. Since the input is a constant, Eq. (6.3) and Eq. (6.5) are equal. The operation of the first-order $\Delta\Sigma$ modulator of Fig. 6.6 is as follows; the input propagates to the quantizer through a integrator and the quantized output is fed back and subtracted from the input signal. This feedback forces the quantized output to track the input and the integrator shapes the quantization error with a high-pass characteristic. A persistent error will accumulate and eventually correct itself. So again, the output is a coarse prediction of the input.

Providing the fractional-N synthesizer with the correct division control is done in the digital domain, meaning that the accumulator and the $\Delta\Sigma$ modulator are both all-digital implementations. Simple digital mapping can equalize their outputs. Due to the digital implementation, matching or non-ideal integrator issues are out of order.

6.4.3 General $\Delta\Sigma$ Modulator Theory

To fully understand the influence of a $\Delta\Sigma$ modulator on the performance of a frequency synthesizer, this section deals with general $\Delta\Sigma$ theory [Nors97]. The presented theory is valid under the assumption that the input signal is sufficiently busy, i.e. the following conditions for the signal and the quantization error are satisfied:

- The quantization error, $e[i]$, is a sample sequence of a stationary random process.

- The input signal changes randomly from sample to sample, such that the error $e[i]$ from sample to sample is largely uncorrelated.

- If the distance between two quantization levels is defined as Δ, the quantization error generated by a busy signal will have equal probability of lying anywhere between $\pm\Delta/2$.

- The random variables of the quantization error process are uncorrelated, i.e. $e[i]$ has a flat power spectral density (it looks like white noise).

The assumption will be referred to as *the white noise assumption*. In its true sense, the assumption never holds, since the quantization error is by definition a deterministic function of the input and hence always statistically dependent on the input. However, the white noise assumption is in many cases a good approximation of reality and permits the use of linear system theory for an intrinsically non-linear system [Nors97]. For a constant input, all the assumption that lead to the white noise assumption are invalid, but the theory can still yield interesting perceptions on the operation of $\Delta\Sigma$ fractional-N synthesizers.

With the white noise assumption, where $e(t)$ has equal probability to lie between $-\Delta/2$ and $+\Delta/2$, the mean square of the quantization error is given by:

$$e_{rms}^2 = \frac{1}{\Delta}\int_{-\Delta/2}^{\Delta/2} e^2 de = \frac{\Delta^2}{12} \tag{6.6}$$

To investigate the influence of the $\Delta\Sigma$ modulator noise on the overall phase noise the PLL bandwidth, the $\Delta\Sigma$ modulator noise in a bandwidth of 1 Hz is of interest. Therefore, the noise spectral density must be calculated which is in contrast to conventional $\Delta\Sigma$ data converter theory, where the *SNR* is more relevant. In this discussion, the noise spectral density is defined at a certain offset from the PLL carrier frequency. Since the noise at both sides of the carrier contributes to the total rms phase error, the noise spectral densities are defined for positive and negative offsets, i.e. double side band. When the noise is sampled at frequency $f_{ref} = 1/T_{ref}$, all of the noise power folds into the frequency band $-f_{ref}/2 \le f < f_{ref}/2$, with copies of this spectrum at each multiple of f_{ref}. The power of the sampled noise spectra is $1/T_{ref}$ times the power of the continuous time signal. Due to the low-pass nature of the PLL, the noise power is restricted to only $-f_c \le f < f_c$, meaning that the relation between continuous and discrete spectra is a simple scaling by $1/T_{ref}$. The spectral density of the quantization error becomes:

$$E(f) = e_{rms}\sqrt{\frac{1}{f_{ref}}} = e_{rms}\sqrt{T_{ref}} \tag{6.7}$$

The following analysis holds for first-order $\Delta\Sigma$ modulators, as shown in Fig. 6.6. The noise added due to quantization is $e[i]-e[i-1]$, i.e. the differentiated error of succeeding quantization steps. The spectral density is then given by :

$$N(f) = E(f)|1 - e^{j\omega T_{ref}}| \tag{6.8}$$

$$= 2e_{rms}\sqrt{T_{ref}}\sin\left(\frac{\omega T_{ref}}{2}\right) \tag{6.9}$$

In the discussion on the total in-band PLL noise, which defines the total rms phase error, the integrated $\Delta\Sigma$ noise power that falls in the PLL noise bandwidth is of importance. The $\Delta\Sigma$ noise power in the PLL band can be calculated as follows:

$$A_{n,\Delta\Sigma}^2 = \int_{-f_c}^{f_c} N^2(f)df \approx e_{rms}^2 \frac{\pi^2}{3} \left(\frac{2f_c}{f_{ref}}\right)^3 \qquad (6.10)$$

where f_c is the cross-over frequency of the PLL and assuming that $f_{ref}^2 \gg f_c^2$. Eq. (6.10) contains the ratio of the sampling frequency f_{ref} and the double of the PLL bandwidth, $2f_c$, which is equivalent to the oversampling ratio in conventional $\Delta\Sigma$ theory, i.e. $OSR = f_{ref}/(2f_c)$.

As can be seen from Eq. (6.10) for a first-order $\Delta\Sigma$ modulator, doubling the oversampling ratio, reduces the in-band noise by 9 dB. This means that if the in-band noise of a $\Delta\Sigma$ fractional-N PLL is determined by the $\Delta\Sigma$ noise, the total phase error can be ameliorated by increasing f_{ref}, while maintaining the same frequency resolution by adding extra bits to the modulator. This is only true if the $\Delta\Sigma$ high-frequency noise is sharply filtered and all components are perfectly linear. In practice however, the $\Delta\Sigma$ noise corrupts the in-band PLL noise through noise leakage by the non-linear mixing in the PD, which will be elaborately discussed later. Even when the OSR is doubled, the in-band noise corruption can remain equal. The real gain from doubling f_{ref} is a decrease in division ratio and hence less in-band noise amplification. Note that a high f_{ref} requires special PD design, leading again to a speed-accuracy trade-off.

6.4.4 $\Delta\Sigma$ Modulators with DC-inputs

From Appendix A, it is obvious that the existence of pattern noise poses a serious problem to the use of $\Delta\Sigma$ modulators in frequency synthesis, where spectral purity is of utmost importance [Gray89] [Cand81][Eynd93]. However, starting from the first-order, single-bit $\Delta\Sigma$ modulator, several improvements can be made to break the patterns in the $\Delta\Sigma$ modulator output, arising from DC-inputs;

- Moving to higher-order $\Delta\Sigma$ modulators is a first solution to remove the pattern noise. Higher-order modulators use improved filter topologies to further reduce the in-band noise generated by the quantization. This is reflected in Eq. (6.11), which is the equivalent of Eq. (6.10) for n-th-order modulators. Doubling the oversampling ratio provides a $3(2n-1)$ dB reduction of the in-band noise.

$$A_{n,\Delta\Sigma}^2 = e_{rms}^2 \frac{\pi^{2n}}{2n+1} \left(\frac{2f_c}{f_{ref}}\right)^{2n+1} \qquad (6.11)$$

  This means that the noise added due to quantization is smaller since higher-order modulators give a better prediction of the input. The randomization provided by higher-order modulators is much better than for first-order $\Delta\Sigma$ modulators, since the predicting action of higher-order $\Delta\Sigma$ modulator is more spread over time, thereby decorrelating the succeeding noise samples. As a result, the *white noise assumption* is much more applicable.

- A second option to improve the prediction made by the modulator and further randomize its output, is multi-bit quantization instead of single-bit. With multi-bit quantization, the output can assume multiple values, such that the modulator prediction is much better and the noise becomes whiter.

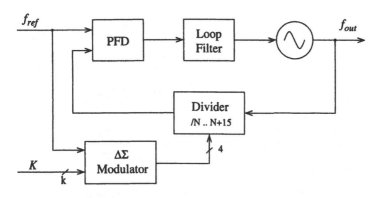

Figure 6.7: *The ΔΣ fractional-N synthesizer*

- Another possibility is to add noise, i.e. *dither*, to the input signal. If properly applied, this noise does not contain energy in the signal band, such that it doesn't degrade the SNR. As a result, the input is no longer a constant and the spurious tones in the output spectrum are smoothed out, but the overall in-band SNR is not substantially decreased. The effects of different amounts and orders of dithering and its effect on ΔΣ fractional-N synthesizers is discussed in more detail later on.

6.5 ΔΣ Modulators for Fractional-N Synthesis

Based on the conclusions, drawn in the previous section, it should be clear that implementing a fractional-N synthesizer with an accumulator type – i.e. a first-order ΔΣ modulator– division control is unwise. These shortcomings are greatly alleviated by implementing the division control with higher-order ΔΣ modulators, which exhibit better randomizing properties, together with lower in-band noise [Mill91, Mill90]. The eventual ΔΣ fractional-N synthesizer architecture is depicted in Fig. 6.7. The ΔΣ modulator maps the k-bit input word onto a 4 bit output word to control the division moduli of the 16-modulus prescaler. The actual principle of fractional division is the same as explained in Section 6.2 and Eq. (6.1) still holds in its raw form.

To investigate the influence of ΔΣ modulators on the spectral purity of the fractional-N synthesizer, this section deals with the design and properties of 3rd-order MASH ΔΣ modulator (cascade 1-1-1) and 3rd-order, multi-bit, single-loop ΔΣ modulators. The MASH and single-loop topologies are chosen as guinea pigs, since they represent the extreme ends of ΔΣ modulator topology spectrum. It goes without saying that other ΔΣ modulator topologies, e.g. cascade 2-1, are equally suitable. They are however not discussed here, since they combine properties of both MASH and single-loop modulators.

To make sure that the ΔΣ modulator noise peak around $f_{ref}/2$ does not corrupt the phase noise of the frequency synthesizer at intermediate offset frequencies, the roll-off of the filtered noise must be equal to or steeper than the roll-off of the PLL phase noise. Since the filter order of the PLL of Chapter 5 is three, the modulator order must be three or higher, as will be explained in

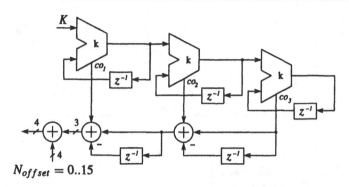

Figure 6.8: *The 3rd-order MASH modulator*

Section 6.6. In this book, only the influence of 3rd-order $\Delta\Sigma$ modulators on the spectral purity of fractional-N frequency synthesizers is investigated in depth, while the effects of higher-order $\Delta\Sigma$ modulators are briefly mentioned.

6.5.1   The MASH Modulator

The MASH (multi-stage noise-shaping) or cascade 1-1-1 $\Delta\Sigma$ modulator is shown in Fig. 6.8. The MASH modulator consists of a cascade of first-order modulators, whose quantization error E_i is the input to the next modulator. By summing the filtered versions of the first-order outputs co_i, the quantization error of the first and the second modulator is cancelled. Since the $\Delta\Sigma$ modulator in fractional-N PLLs is an all digital implementation, the cancellation is perfect. The output of the modulator with perfect cancellation is:

$$Y = co_1 + (1 - z^{-1}) \cdot co_2 + (1 - z^{-1})^2 \cdot co_3$$
$$= K + (1 - z^{-1})^3 \cdot E_3 \tag{6.12}$$

The equations expose the most important quality of a MASH modulator, i.e. its unconditional stability for any modulator order, because of its first-order nature. Another advantage is the integratability of the MASH modulator in plain CMOS technology, since only adders and registers are needed to implement the noise shaping function.

In the implementation of Fig. 6.8, the output is a 3-bit word with a mean value n, which is a fractional number between 0 and 1 with k-bit accuracy. In order to synthesize the total frequency range of the PLL, a 4-bit word, N_{offset}, is added to the output, choosing the proper integer division modulus. The resulting 4-bit word controls the prescaler moduli. To increase the output dynamic range of the modulator to accommodate more division moduli, multiple MSBs can be taken as outputs of the first-order modulators instead of the single bit output, co_i.

The MASH modulator exploits its full dynamic range to come to the wanted mean output value, i.e. it employs all possible output states, 8 in this case. In the time domain, this is reflected in the intensive prescaler modulus switching. To synthesize for instance a frequency of $67.92 \times f_{ref}$, i.e. $N_{offset} = 3$ and $K = 0.92 \cdot 2^k$, all moduli between 64 and 71 are employed, as

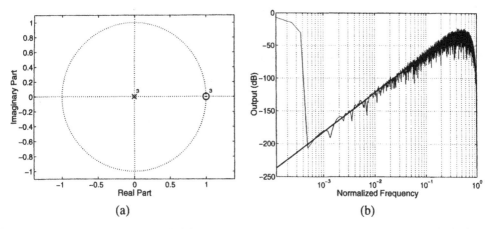

Figure 6.9: *(a) The pole-zero plot of the 3rd-order MASH modulator and (b) the theoretical (solid line) and simulated output spectrum for a frequency of 67.92 × f_{ref}.*

shown in Fig. 6.10. As a consequence, the dynamic range of the division modulus is limited by the modulator. To synthesize the intended DCS-1800 band, the necessary dynamic range is 9 centered around 69 for $f_{ref} = 26$ MHz. The attainable dynamic range with a MASH modulator is between 67 and 76, i.e. 9, but centered around 71.5. To achieve the wanted frequency band the reference frequency should be changed to 25 MHz or the modulus range of the prescaler should be extended.

In the frequency domain, the intensive use of modulus dynamic range translates in substantial levels of high frequency ΔΣ noise. This is reflected in the Noise Transfer Function (NTF) of the MASH modulator, which is $H_{qn}(z) = (1 - z^{-1})^3$ (see Eq. (6.12)). The NTF contains 3 poles at the origin of the z-plane and 3 high-pass zeros at the unity circle (see Fig. 6.9 (a)). In Fig. 6.9 (b), the theoretical PSD (Power Spectral Density) is plotted together with the FFT of the simulated MASH modulator output. The PSD can be calculated from the NTF by substituting $z = e^{j2\pi f/f_{ref}}$ and knowing that the PSD of the quantization noise is given by $\Delta^2/(12 \cdot f_{ref})$ (see Eq. (6.7)):

$$S_f(f) = \frac{\Delta^2}{12 \cdot f_{ref}} \left[2 \sin\left(\frac{\pi f}{f_{ref}}\right) \right]^{2n} \tag{6.13}$$

with n the order of the ΔΣ modulator, in this case 3, and the quantization step $\Delta = 1$, which is discussed later.

6.5.2   The Multi-Bit, Single-Loop ΔΣ Modulator

The multi-bit, single-loop ΔΣ modulator is shown in Fig. 6.11. In contrast to the MASH modulator, this modulator consists of a single, 3rd-order discrete time filter with feedforward and feedback coefficients, which influence the NTF. The value of the coefficients is derived from a 3rd-order, high-pass Butterworth filter implementation. The cut-off frequency of the Butterworth

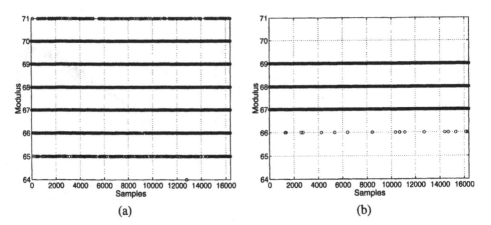

Figure 6.10: *The time domain representation of the division moduli to synthesize a frequency of* 67.92 × f_{ref}, *employed by (a) the 3rd-order MASH modulator and (b) the multi-bit, single-loop modulator.*

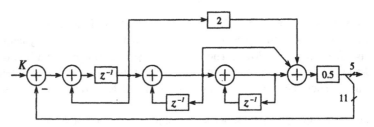

Figure 6.11: *The 3rd-order multi-bit, single-loop $\Delta\Sigma$ modulator*

filter is chosen to be sufficiently lower than $f_{ref}/2$ and such that the coefficients are close to powers of two. The implemented filter has a cut-off frequency of 0.167 × f_{ref}, leading to the noise transfer function of Eq. (6.14).

$$H_{qn,b}(z) = \frac{(1-z^{-1})^3}{1 - 0.968z^{-1} + 0.587z^{-2} - 0.106z^{-3}} \quad (6.14)$$

To enable integration of the digital filter in plain CMOS technology without implementing multipliers, the coefficients in the denominator are approximated such that the implemented feedforward and feedback coefficients are power of two. The resulting NTF is given in Eq. (6.15), which is close to the Butterworth NTF, with preservation of the stability and causality conditions.

$$H_{qn}(z) = \frac{(1-z^{-1})^3}{1 - z^{-1} + 0.5z^{-2}} \quad (6.15)$$

In contrast to the MASH modulator, the NTF contains only 1 pole at the origin of the z-plane and two low-Q Butterworth poles at 0.167 × f_{ref}. In Fig. 6.12 (a), the poles and zeros are plotted in the

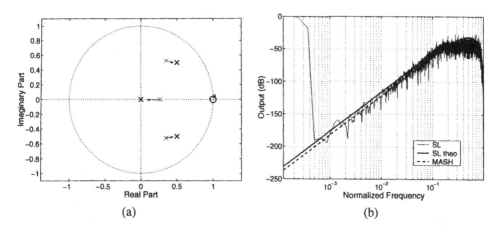

Figure 6.12: *(a) The pole-zero plot of the 3rd-order multi-bit, single-loop modulator together with the poles of the theoretical Butterworth transfer function (light crosses) and (b) the theoretical (solid line) and simulated output spectrum for a frequency of* 67.92 × f_{ref}*, compared to the simulated output spectrum of the MASH modulator (dashed line).*

z-plane. The arrows indicate the alteration of the pole positioning due to the approximation of the Butterworth filter transfer function. In Fig. 6.12 (b), the output of the modulator is compared to its theoretical PSD and the PSD of the MASH modulator. Although the single-loop ΔΣ modulator is more complex than the MASH modulator, it offers a much higher flexibility in terms of noise shaping. The quantization noise in the passband of the modulator is smoothed out, by proper pole positioning. The passband gain is only 3.2 compared to 8 for the MASH modulator. As a result, the HF ΔΣ noise is less likely to corrupt the PLL phase noise. Note that the input is also subjected to a non-unity transfer function $H_{in}(z)$ given in Eq. (6.16). This poses no problems for frequency synthesis, since the DC value of the input is passed unaltered. But problems arise when the synthesizer is used for data transmission, where the modulated data stream will be shaped by the ΔΣ modulator signal transfer function.

$$H_{in}(z) = \frac{z^{-1}(4 - 5z^{-1} + 2z^{-2})}{2 - 2z^{-1} + z^{-2}} \tag{6.16}$$

As discussed, the NTF of the multi-bit, single-loop ΔΣ modulator is designed to have less HF noise. In the time-domain, this translates to less intensive prescaler modulus switching as can be seen in Fig. 6.10 (b). Only the moduli between 66 and 69 are needed to synthesize a frequency of 67.92 × f_{ref}. In fact, a qualitative relation with the passband gain can be established. For a MASH modulator, the passband gain is 8 and 8 output states are employed to obtain the wanted output. For the single-loop modulator, the passband gain is 3.2 and 3 output states are intensively used while the fourth output state is used only for limited amount of the time. Although the use of output states is also dependent on the input, it can be concluded that by lowering the passband gain, the prescaler modulus switching is reduced.

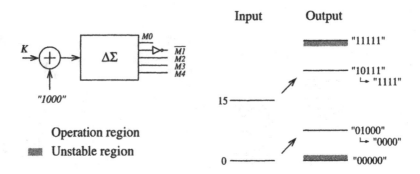

Figure 6.13: *The transformation of the input range of the multi-bit ΔΣ modulator to a 5-bit output to avoid overlap with unstable regions. All input numbers are normalized to 2^k.*

In the ΔΣ modulator of Fig. 6.11, the output is scaled by 0.5 before quantization for stability reasons. Different implementations of the same $H_{qn}(z)$ of Eq. (6.15) are possible, but this implementation results in coefficients {2,1,0.5} which are easy to integrate. Due to the all digital implementation no DAC needs to be implemented in the feedback of the remaining LSBs, while the stability of the modulator is greatly enhanced by the multi-bit approach. In spite of the stability enhancement an input region for which the ΔΣ modulator becomes unstable exists at the edges of the input range. Therefore, a 5-bit quantization is applied, by merely isolating the 5 MSBs, although only 4 output bits are needed to control the prescaler moduli. By offsetting the input by "1000", the needed 4-bit output range is located in the middle of the total output range. The resulting output bits are mapped to their 4-bit equivalent by taking only the last 4 of the 5 MSBs {$M1, M2, M3, M4$} and by inverting $M1$, i.e. a subtraction by 8. Fig. 6.13 demonstrates this technique. As can be seen, the unstable input regions are avoided.

The multi-bit, single-loop modulator addressed only a fraction of its full dynamic output range to come to the wanted mean value. As a consequence, the dynamic range of the division modulus is enhanced compared to the MASH modulator. The attainable dynamic range is now from 65 to 78, i.e. 12 centered around 71.5, which is sufficient to synthesize the full DCS-1800 band with a reference frequency of 26 MHz. Note that the edges of the modulus range (i.e. 64 and 79) are not usable, alleviating the need for the 5-bit quantization, since the unstable input regions are much smaller than a LSB due to the multi-bit approach. In this design case a 4-bit quantization would be satisfactory, but the 5-bit case is discussed for the sake of completeness.

6.6   The Theoretical ΔΣ Phase Noise Analysis

In this section, a theoretical analysis of the impact of ΔΣ noise on the spectral purity of the frequency synthesizer is worked out. The analysis is split up in two parts: out-of-band ΔΣ phase noise and in-band ΔΣ phase noise, since both noise contributions affect different synthesizer specifications. For the analysis, a linear-time-invariant (LTI) PLL model is employed to calculate he $\Delta\Sigma$ noise as it appears at the output of the PLL. This means that the effect of tonal energy n the modulator spectrum, transient effects and the influence of the non-linearities are not taken nto account.

5.6.1 The Out-of-Band $\Delta\Sigma$ Phase Noise

For compatibility with the LTI PLL model of the previous chapter, the $\Delta\Sigma$ noise is modeled as an additive noise source $\theta_{\Delta\Sigma}(s)$ at the input of the prescaler. Since the prescaler and the $\Delta\Sigma$ modulator are digital, sampled systems, the modeling effort is performed in the z-domain. To ind the influence of the $\Delta\Sigma$ noise on the phase within the loop, the prescaler-$\Delta\Sigma$ modulator combination must be examined in detail; The $\Delta\Sigma$ modulator output bits control the pulse swallowing action of the prescaler, such that a mean division by $N + K/2^k$ is obtained. In fact, the prescaler with $\Delta\Sigma$-control can be looked upon as a digital-to-phase converter (DPC). Similar to a digital-to-analog converter (DAC), where the digital input controls the analog output amplitude, the digital input now controls the output phase/frequency of the prescaler. Every reference cycle, the prescaler subtracts $i \cdot 2\pi$ rad from its input phase, with $i = 0 .. 2^b - 1$ determined by the b-bit $\Delta\Sigma$ modulator output. In other words, the division modulus is modulated by the $\Delta\Sigma$ output, by the wanted mean value as well as by the shaped high frequency noise, leading to Eq. (6.17) with $E(z)$ the quantization noise.

$$N(z) = N + \frac{K}{2^k} + H_{qn}(z) \cdot E(z) \tag{6.17}$$

Since the $\Delta\Sigma$ modulator modulates the division modulus, the output frequency of the PLL is also modulated by the $\Delta\Sigma$ modulator. The output frequency of the PLL is given in Eq. (6.18). The first term is the wanted output frequency, determined by the mean $\Delta\Sigma$ output. Since this is a constant, i.e. DC value, it passes unaltered through the PLL. The second term consists of the frequency fluctuations due to the $\Delta\Sigma$ quantization noise, which are subjected to the closed loop transfer function of the PLL, $T(z)$.

$$f_{out} = \left(N + \frac{K}{2^k}\right) \cdot f_{ref} + T(z) \cdot H_{qn}(z) \cdot E(z) \cdot f_{ref} \tag{6.18}$$

To determine the influence on spectral purity, the PSD of the frequency fluctuations at the input of the prescaler is calculated in Eq. (6.19).

$$S_f(z) = |H_{qn}(z) \cdot f_{ref}|^2 \cdot \frac{\Delta^2}{12 \cdot f_{ref}} \tag{6.19}$$

As elaborately discussed in Section 6.4.3, the quantization noise is approximated by uniformly distributed white noise, although the input to the $\Delta\Sigma$ is constant. The quantization noise power is then $\Delta^2/12$. The quantization step Δ is determined by the smallest possible phase step the $\Delta\Sigma$ prescaler can handle. In this case, the minimal phase step is 2π rad of the prescaler input signal, which is equivalent to a unity change of the prescaler modulus N. The maximum phase step range is given by the prescaler modulus range $\Delta N = N_{max} - N_{min}$, which is 15 in the

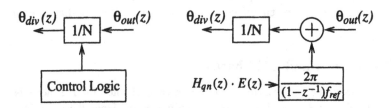

Figure 6.14: *The equivalent model for the prescaler with $\Delta\Sigma$ modulator.*

design case. The number of quantization steps of the $\Delta\Sigma$ modulator is given by the number of effective output bits b, which is 4 for both modulator types. The resulting quantization step Δ is then $\Delta N/(2^b - 1) = 1$ for both modulator types. This could be intuitively seen, since the only possible minimal phase change in the synthesizer loop is unity change of the division modulus, which has always the same size, regardless of the number of modulator output bits or the modulus range. The PSD of the "whitened" quantization noise is then $1/(12 \cdot f_{ref})$.

By a rectangular integration of the PSD of the frequency fluctuations in the z-domain, the phase noise contribution of the $\Delta\Sigma$ modulator is found as an additive noise source at the input of the prescaler:

$$S_{\theta,\Delta\Sigma}(z) = \frac{|H_{qn}(z)|^2 \cdot f_{ref}}{12} \cdot \frac{(2\pi)^2}{|1 - z^{-1}|^2 \cdot f_{ref}^2}$$

$$= \frac{\pi^2}{3 \cdot f_{ref}} \cdot \frac{|H_{qn}(z)|^2}{|1 - z^{-1}|^2} \tag{6.20}$$

In short, the $\Delta\Sigma$ noise is converted to phase noise by an integration in the prescaler, which could be expected since the prescaler divides frequency, not phase. In the LTI model, the prescaler can be replaced by the model given in Fig. 6.14.

With Eq. (6.20) the influence of the $\Delta\Sigma$ noise on the spectral purity of the frequency synthesizer can be examined. In Fig. 6.15, the $\Delta\Sigma$ of the MASH and the single-loop modulator, as it appears at the synthesizer output is plotted versus the simulated (Matlab) output phase noise from the LTI model in Chapter 5 (dashed-dotted curve). Obviously, the $\Delta\Sigma$ phase noise poses no threat to the spectral purity. To ensure that the $\Delta\Sigma$ noise peak does not corrupt the output phase noise at intermediate offset frequencies (600 kHz to 10 MHz), the order of the $\Delta\Sigma$ modulator needs to be carefully chosen. For simplicity, the calculation of the necessary order of the $\Delta\Sigma$ modulator is performed for a MASH modulator of order n, i.e. $H_{qn}(z) = (1 - z^{-1})^n$. The PLL transfer function is approximated by an m-th-order Butterworth filter, i.e. $T(f) = \left(1 + j\,(f/f_c)^m\right)^{-1}$, to have a $-m \cdot 20$ dB/dec roll-off at the frequencies of interest. The $\Delta\Sigma$ output phase noise becomes then:

$$S_{\theta,\Delta\Sigma,o}(f) = \left[\frac{(2\pi)^2}{12 \cdot f_{ref}} \left[2\sin\left(\frac{\pi f}{f_{ref}}\right) \right]^{2(n-1)} \right] \cdot |T(f)|^2 . \tag{6.21}$$

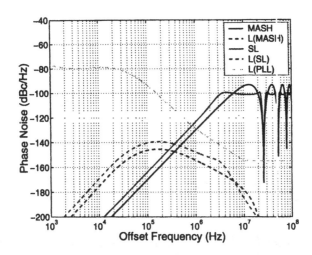

Figure 6.15: *The results of the theoretical analysis compared to the simulated PLL output phase noise (dash-dotted line). The dashed lines represent the noise of both $\Delta\Sigma$ modulators, as it appears at the PLL output.*

Since the offset frequencies of interest are $f_c \ll f < f_{ref}$, the following approximations can be made: $\sin(\pi f/f_{ref}) \cong (\pi f/f_{ref})$ and $|T(f)|^2 \cong (f_c/f)^{2m}$. By applying these approximations to Eq. (6.21) and by converting $S_{\theta,\Delta\Sigma,o}$ to dBs, the simplified expression of Eq. (6.22) is found for the $\Delta\Sigma$ phase noise at the PLL output.

$$10\log(S_{\theta,\Delta\Sigma,o}(f)) \approx 10\log\left(\frac{(2\pi)^{2n}}{12}\left(\frac{1}{f_{ref}}\right)^{2n-1} f_c^{2m}\right) + 20\,(n-1-m)\log f \qquad (6.22)$$

The roll-off of the $\Delta\Sigma$ noise at the output of the synthesizer is $-20\,(m-n+1)$ dB/dec. At these offset frequencies the phase noise of the PLL has a roll-off of -20 dB/dec, due to the white noise determined phase noise in the VCO. To prevent corruption of the output phase noise by the $\Delta\Sigma$ noise, the roll-off of the $\Delta\Sigma$ phase noise should be equal or higher than the roll-off of the loop filter transfer function, i.e. $m \geq n$. Since the order of the loop filter is three in the presented design case (see Chapter 5), the $\Delta\Sigma$ modulator order must be three or lower. It goes without saying, that this statement is also valid for the order of multi-bit, single-loop $\Delta\Sigma$ modulators in fractional-N frequency synthesizers.

Since the main advantage of $\Delta\Sigma$ fractional-N synthesizers is the decoupling of the choice of the reference frequency f_{ref} and the PLL bandwidth f_c, the influence of the $\Delta\Sigma$ noise on the bandwidth requirement is examined. From Eq. (6.21) and Eq. (6.22), the maximum bandwidth $f_{c,max}$ of the PLL can be derived such that the $\Delta\Sigma$ produced phase noise does not corrupt the wanted phase noise specification for a given reference frequency f_{ref}. The target phase noise spec is set at 3 MHz for two reasons; First, the $\Delta\Sigma$ phase noise is closest to the PLL output phase noise at around 3 MHz (see Fig. 6.15), making it the most crucial offset frequency. Secondly,

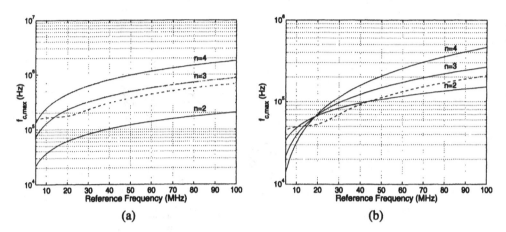

(a)   (b)

Figure 6.16: *The maximum PLL bandwidth $f_{c,max}$ versus the reference frequency and different ΔΣ modulator orders, equal to the PLL loop filter order for (a) the Butterworth approximation and (b) the type-II, 4th-order PLL of Chapter 5. The dashed curve is for the 3rd-order, single-loop modulator. The dashed curve is for the 3rd-order, single-loop modulator. The targeted phase noise specification is -136 dBc/Hz at 3 MHz for DCS-1800.*

the blocker specification of the DCS-1800 poses the most stringent demands on the phase noise at this offset frequency, i.e.-133 dBc/Hz. To have some margin on the PLL noise, the noise spec for the ΔΣ is set to $S_{\theta,\Delta\Sigma,o}(3 \text{ MHz}) = -136$ dBc/Hz. The PLL transfer function is assumed to be a Butterworth filter, with its order equal to the order of the ΔΣ modulator, $m = n$. From Eq. (6.22), the maximum bandwidth requirement can be calculated for an n-th-order MASH modulator to be:

$$f_{c,max} = \left[S_{\theta,\Delta\Sigma,o}(f) \cdot \frac{12}{(2\pi)^{2n}} \cdot f_{ref}^{2n-1} \cdot f^2 \right]^{\frac{1}{2n}} \tag{6.23}$$

In Fig. 6.16, the $f_{c,max}$ is plotted versus the reference frequency for different ΔΣ modulator orders, equal to the order of the PLL loop filter. The dashed curve represents $f_{c,max}$ for the single-loop, multi-bit ΔΣ modulator of Section 6.5.2. The transfer function of Eq. (6.15) can be seen as that of the MASH modulator, with a factor 4 increase in noise power due to the denominator of the transfer function. This approximation holds for offset frequencies up to 1 MHz. Since the frequency range of interest is up to 3 MHz, the actual transfer function is used to calculate the maximum bandwidth. Also for reference frequencies lower than 18 MHz, the above approximations lose their validity (see the flat region for the single-loop ΔΣ modulator). Fig. 6.16 shows that for increasing f_{ref} and increasing order, the maximum possible bandwidth increases, enabling faster PLL dynamics, without ΔΣ noise degradation. However, the price to be paid is a higher system complexity for higher orders and a significant increase in power consumption for higher reference frequencies. For the design case, f_{ref} is 26 MHz and the modulator and filter order is three. The maximum PLL bandwidth is then 284 kHz for the MASH modulator and 204 kHz for the single-loop, multi-bit modulator, which is approximately a factor 100 smaller than the reference frequency. If the $f_{c,max}$ is calculated for the other important DCS-1800 phase noise spec, i.e.-119 dBc/Hz at 600 kHz, the resulting maximum bandwidth is much higher: 319 kHz and 251 kHz, respectively. Note that for reference frequencies lower than approximately 13.8 MHz, the single-loop modulator yields a higher maximum bandwidth than the MASH modulator, due to the pole shift in the NTF.

In Fig. 6.16 (b), the $f_{c,max}$ is plotted versus the reference frequency for different $\Delta\Sigma$ modulator orders but for the type-II, 4th-order PLL of Chapter 5. The transfer function $T(f)$ of the PLL can no longer be approximated by the simple Butterworth filter with three poles at f_c. By substituting the frequency of the zero by $\alpha^{-1}f_c$, the pole frequencies by βf_c and C_z by Eq. (5.16) in the open-loop transfer function $GH(f)$ of the PLL (Eq. (5.14)), the open-loop gain can be approximated by $\beta^2 (f_c/f)^3$ in the frequency range $f_c \ll f < f_{ref}$. The transfer function of the PLL becomes then:

$$|T(f)|^2 = \frac{\beta^4 f_c^6}{f^6 + \beta^4 f_c^6 + 2\beta^2 f_c^3} \cong \beta^4 \frac{f_c^6}{f^6} \qquad (6.24)$$

The transfer function of Eq. (6.24), is in fact an approximation of a 3rd-order Butterworth filter with one pole at f_c and two poles at βf_c. Substituting the $|T(f)|^2$ in Eq. (6.21) for the frequency range of interest leads to the formula for the maximum bandwidth of the PLL:

$$f_{c,max} = \left[S_{\theta,\Delta\Sigma,o}(f) \cdot \frac{12}{(2\pi)^{2n}} \cdot f_{ref}^{2n-1} \cdot \frac{f^{8-2n}}{\beta^4} \right]^{1/6} \qquad (6.25)$$

The result is plotted in Fig. 6.16 (b). The dashed line is again the maximum bandwidth for the single-loop, multi-bit $\Delta\Sigma$ modulator. For a reference frequency of 26 MHz, not much is gained from increasing the modulator order. For a high bandwidth and thus a fast PLL, the reference frequency should be increased leading to an increased power consumption. Note that for f_{ref} below 20 MHz, a higher $\Delta\Sigma$ order leads to a lower possible PLL bandwidth! The maximum bandwidth is 87 kHz for the 3rd-order MASH modulator and 62 kHz for the single-loop, multi-bit modulator. This is significantly lower than would be expected from Fig. 6.16, due to the shifting of two poles towards higher frequencies with a factor $\beta = 6$ to ensure enough phase margin in the phase-locked loop.

6.6.2 The $\Delta\Sigma$ rms Phase Error

Besides the out-of-band phase noise, the in-band phase noise due to the $\Delta\Sigma$ modulation is of importance, since it contributes to the rms phase error, $\Delta\Phi_{rms}$, of the synthesizer.

Similar to DACs, the dynamic range of the fractional-N synthesizer can be defined. The dynamic range of a synthesizer is the ratio of the largest possible frequency amplitude (i.e. change) and the smallest one. The largest frequency change is the full frequency range of the modulator $\Delta N \cdot f_{ref}$ with ΔN the modulus range. The smallest possible frequency change is limited by the in-band phase noise. The in-band phase noise gives rise to an equivalent frequency noise Δf_n, which sets the accuracy of the frequency generation. If the in-band phase noise is $10 \log A_n$ in dBc/Hz, the equivalent frequency noise can be calculated to be:

$$\Delta f_n = \sqrt{\int_{-f_{nbw}}^{f_{nbw}} A_n f^2 df} = \sqrt{\frac{2}{3} A_n f_{nbw}^3}$$ (6.26)

with f_{nbw} the noise bandwidth of the frequency synthesizer such that the phase noise equals A_n for $f \le f_{nbw}$ and 0 for $f > f_{nbw}$. The noise bandwidth of a first-order system with a bandwidth f_c is $\frac{\pi}{2} f_c$, while the noise bandwidth of systems with an order higher than five is equal to f_c. The noise bandwidth of the PLL of Chapter 5 is $f_{nbw} \approx 1.16 f_c$. The relation between the in-band noise A_n and the rms phase error is:

$$A_n = \frac{\Delta\Phi_{rms}^2}{2} f_{nbw}^{-1}$$ (6.27)

Using Eq. (6.26) and Eq. (6.27), the frequency noise as a function of the rms phase error for the 4th-order PLL is calculated in Eq. (6.28).

$$\Delta f_n = \sqrt{\frac{2}{3} \frac{\Delta\Phi_{rms}^2}{2} f_{nbw}^2} \cong \sqrt{\frac{\Delta\Phi_{rms}^2}{2} f_c^2}$$ (6.28)

From Eq. (6.28), the dynamic range of the frequency synthesizer can be derived:

$$DR_{PLL} = \left(\frac{\Delta N \cdot f_{ref}}{\Delta f_n}\right)^2 = 8 \cdot \frac{\Delta N^2}{\Delta\Phi_{rms}^2} \cdot \left(\frac{f_{ref}}{2 f_c}\right)^2$$ (6.29)

Knowing the dynamic range of the frequency synthesizer, the necessary specifications of the $\Delta\Sigma$ modulator in terms of phase error can be derived; First, the dynamic range of an n-th-order MASH modulator is given in Eq. (6.30) [Nors97] with n the $\Delta\Sigma$ modulator order, b the number of output bits of the modulator, determining the largest possible output signal:

$$DR_{\Delta\Sigma} = \frac{3}{2} \cdot (2^b - 1)^2 \cdot \frac{2n+1}{\pi^{2n}} \cdot \left(\frac{f_{ref}}{2 f_c}\right)^{2n+1}$$ (6.30)

To make sure that the $\Delta\Sigma$ modulator does not corrupt the rms phase error, the dynamic range of the $\Delta\Sigma$ modulator must be higher than the dynamic range of the frequency synthesizer. Using the above expression for the respective dynamic ranges and knowing that $(2^b - 1)^2 = \Delta N^2$, the inequality can be rewritten to:

$$\frac{3}{2} \cdot \frac{2n+1}{\pi^{2n}} \cdot \left(\frac{f_{ref}}{2 f_c}\right)^{2n+1} > \frac{8}{\Delta\Phi_{rms}^2} \cdot \left(\frac{f_{ref}}{2 f_c}\right)^2$$ (6.31)

After some straightforward math, the maximum bandwidth of the PLL for which the $\Delta\Sigma$ noise does not corrupt the rms phase error is calculated in Eq. (6.32).

$$f_c < \left[\frac{3}{8} \cdot \frac{2n+1}{(2\pi)^{2n}} \cdot \Delta\Phi_{rms}^2\right]^{\frac{1}{2n-1}} \cdot f_{ref}$$ (6.32)

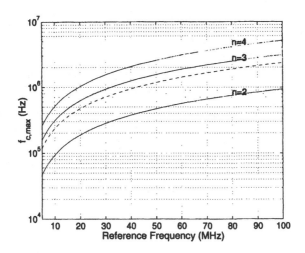

Figure 6.17: *The maximum PLL bandwidth $f_{c,max}$ versus the reference frequency and different $\Delta\Sigma$ modulator orders for $\Delta\Phi_{rms} < 1.5^\circ$. The dashed curve is for the 3rd-order, single-loop modulator.*

With these formulae, the maximum PLL bandwidth $f_{c,max}$ is plotted versus the reference frequency of the PLL in Fig. 6.17. The plotted results only hold for n-th-order MASH modulators. For the single-loop, multi-bit $\Delta\Sigma$ modulators, the actual maximum bandwidth can be calculated to be 25% smaller than in Eq. (6.32). Due to the Butterworth poles, the in-band quantization noise power of the single-loop modulator is 4 times higher than that of a 3rd-order MASH converter. The figure shows that the maximum bandwidth of the PLL increases with increasing f_{ref} and increasing modulator order, as could be expected. In the case of a 3rd-order modulator, a 1.5° rms phase error and a f_{ref} of 26 MHz, the maximum bandwidth is 810 kHz and 614 kHz, respectively. Even for a rms phase error of only 1°, $f_{c,max}$ is still 690 kHz and 522 kHz. Clearly, the constraints posed on the $\Delta\Sigma$ modulator noise due to in-band noise contributions are much less severe than the constraints due to the out-of-band phase noise at 3 MHz (see the previous section). This means that the $\Delta\Sigma$ modulator does not corrupt the rms phase error of the PLL.

6.7 A Fast Non-Linear $\Delta\Sigma$ Phase Noise Analysis Method

The theoretical analysis suggests that applying $\Delta\Sigma$ control to the prescaler would not cause any major problems for the spectral purity of the PLL. The out-of-band $\Delta\Sigma$ produced phase noise seems to be the bottleneck in $\Delta\Sigma$ fractional-N synthesizer design for low phase noise. Practice however proves this wrong; It is a well known fact that fractional-N synthesizers suffer from spurious tones. What's more, since the $\Delta\Sigma$ fractional-N synthesizer can be viewed as a $\Delta\Sigma$ Digital-to-Frequency converter (DFC), noise leakage due to non-linearities in the conversion is bound to degrade the dynamic range of the synthesizer, as is the case in $\Delta\Sigma$ DACs [Nors97].

The proposed theoretical analysis (Section 6.6) is based on a linear time-invariant PLL model.

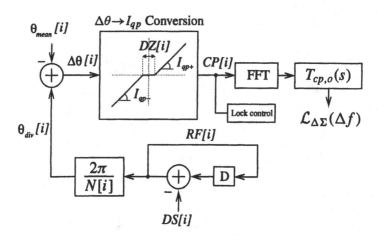

Figure 6.18: *The schematic representation of the non-linear, transient analysis method with all corresponding simulation parameters and variables. The digital signal path of the fractional-N synthesizer is simulated in open loop and in discrete time.*

For phase noise analysis, the PLL is assumed to be in a steady operating point, i.e. in lock, around which the linear assumption is valid. The analysis is not capable of incorporating any time effects and their effect on the spectral purity of the PLL. Since a fractional-N synthesizer switches continuously between different operating points, it is in fact never in lock and the LTI analysis is not applicable. Although the theoretical $\Delta\Sigma$ phase noise analysis can provide interesting insights, an analysis method is mandatory that takes into account time effects and non-linearities in the PLL building blocks.

6.7.1   The Analysis Method

In this book, an analysis method is developed, that provides a fast, interactive means of examining the influence of non-linearities and mismatch in different PLL building blocks by performing a transient simulation of the $\Delta\Sigma$ fractional-N synthesizer. The strong point of the analysis method is its capability to sweep simulations sufficiently fast over different degrees of non-linearities and operating points, while performing sufficiently long transient simulations to get accurate fast-Fourier transforms (FFT) of the phase variable. This is accomplished by analyzing the fractional operation of the PLL in discrete time and in open-loop. Since all signal processing from the VCO output to the filter input is sampled, a discrete time analysis can be employed. All sampling, except in the prescaler, is assumed to be at the edges of the reference frequency signal. To further speed up the simulation, the building blocks are represented by high level models with parameters to model any non-linear behavior or mismatch in critical transistors. Note that jitter on the reference clock and thus on the $\Delta\Sigma$-generated modulus control pulses is not incorporated in the analysis.

The simulated signal path with the different simulation variables and parameters is presented in Fig. 6.18. The analysis is implemented in MATLAB [Mat97]. The lock control is implemented such as to ensure closed loop conditions, i.e. that the phase and voltage excursions have a mean value that corresponds to the wanted output frequency. As can be seen in Fig. 6.18, the simulation starts with the variable $RF[i]$, which represents the accumulated variation of the number of RF pulses, generated by the VCO. Every reference cycle, the number of RF pulses at the divider output is determined by the number of input RF pulses swallowed by the $\Delta\Sigma$ control. The $\Delta\Sigma$ output, $DS[i]$ is obtained from transient simulations of the respective $\Delta\Sigma$ modulators of Section 6.5.1 and Section 6.5.2. The resulting state changes of the $\Delta\Sigma$ modulator outputs can be derived from Fig. 6.10. The variation of the number of RF pulses at time $[i]$ is given by the number of RF pulses at the previous reference cycle minus the number of pulses swallowed by the prescaler due to the $\Delta\Sigma$ control:

$$RF[i] = \sum_{k=2}^{i} RF[k-1] - DS[k] \tag{6.33}$$

The $\Delta\Sigma$ control changes the number of swallowed RF pulses at every reference cycle, giving rise to a sudden phase switch at the output of the prescaler. Consequently, a phase error is generated every reference cycle, meaning that the loop is almost never phase locked. To find the phase error, the phase changes need to be compared to the phase that would be expected when the loop would be in phase lock. This is the mean phase, $\theta_{mean}[i]$, which corresponds to the wanted fractional part of the division modulus, $K/2^k$. The accumulated phase error at a given moment $[i]$, $\Delta\theta[i]$, is given in Eq. (6.34). $N[i]$ is the instantaneous division modulus.

$$\Delta\theta[i] = \sum_{k=2}^{i} \frac{2\pi}{N[k]} \left(RF[k] - (k-1) \cdot \frac{K}{2^k} \right) \tag{6.34}$$

The phase error is converted to current pulses, $CP[i]$, in the phase-frequency detector and the charge pump (see Eq. (6.35)). The $\Delta\theta \rightarrow I_{qp}$ (phase error to charge pump current) conversion is modeled to contain different degrees of non-idealities of the actual charge pump and PFD circuit integration. Mismatch in the up and down current sources, resulting in gain mismatch for positive and negative phase errors is modeled by $I_{qp\pm}$. The occurrence of a hard non-linearity around zero phase error, i.e. a dead zone, is modeled by $DZ[k]$. The size of the dead zone is expressed in percents of the useful $\Delta\theta \rightarrow I_{qp}$ conversion range, i.e. -2π to 2π in this design case. The circuit implications of the non-idealities are discussed in Section 6.8.

$$CP[i] = \sum_{k=2}^{i} \frac{I_{qp\pm}}{2\pi} (\Delta\theta[k] \pm DZ[k]) \tag{6.35}$$

By taking a fast Fourier transform of the current pulses, the current noise spectrum is obtained. The current noise spectrum is modeled as a noise source in the LTI PLL model of Chapter 5. To find the output phase noise the current noise is subjected to its corresponding closed loop transfer function $T_{cp,o}(s)$. Any mismatches in the R's, C's and loop filter path addition can

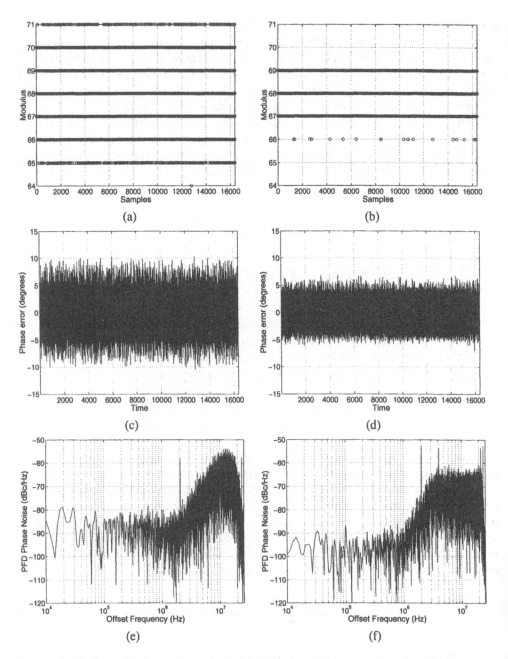

Figure 6.19: *Simulation results: The division moduli of (a) the MASH modulator and (b) the single-loop, multi-bit modulator. The phase error* $\Delta\theta$ *for (c) the MASH modulator and (d) the single-loop, multi-bit modulator. The FFT of the current pulses* $CP[i]$ *for (e) the MASH modulator and (f) the single-loop, multi-bit modulator.*

e modeled using $\Delta\alpha$, $\Delta\beta_1$ and $\Delta\beta_2$, for the uncertainty on the zero position and the positions f poles τ_p and τ_4, respectively (see Eq. (6.36) and Chapter 5).

$$T_{cp,o}(s) = \frac{K_{vco}(s)N\left(1 + \frac{(\alpha\pm\Delta\alpha)s}{\omega_c}\right)}{s^2 C_z N\left(1 + \frac{s}{(\beta_1\pm\Delta\beta_1)\omega_c}\right)\left(1 + \frac{s}{(\beta_2\pm\Delta\beta_2)\omega_c}\right) + K_{pd}(s)K_{vco}(s)\left(1 + \frac{(\alpha\pm\Delta\alpha)s}{\omega_c}\right)}$$

$$(6.36)$$

With this linearized approach, the speed and transparency of the simulation is maintained. The on-linear conversion from voltage to frequency/phase in the VCO is modeled by the variation f the VCO gain $K_{vco}(f)$, when changing the operating point of the PLL. The tuning of the VCO as an almost constant second derivative, such that the K_{vco} can be approximated by a first-order unction of the frequency:

$$K_{vco}(f) = 2\pi(-0.89 \cdot f + 1.7815e9) \tag{6.37}$$

Due to the equalization in the PLL to keep $K_{vco} \times I_{qp}$ constant, the PFD gain changes with the PLL operating point and is therefore also a function of the output frequency of the synthesizer, i.e. $K_{pd}(f)$. The approach is sound because the phase variations due to the fractional operation of he PLL result in very small voltage excursions at the output of the filter. Moreover, the transfer characteristics of the VCO and the loop filter are monotonic and mildly non-linear, justifying a linearization around their operating points. The linearization has minor implications on the accuracy of the simulation, since the non-linearities introduced by the filter and the VCO have small impact with respect to the non-linearity of the PFD-charge pump circuit (see Section 6.8). To increase the accuracy of the simulator, the non-linear characteristic of the VCO tuning diodes must be included. Therefore, a full transient simulation of the fractional-N PLL in closed-loop s mandatory, resulting in lengthy simulations due the high output frequency of the VCO. The presented open-loop, linearized approach enables a full non-linear analysis in approximately 15 seconds on a Pentium4, including the transient simulation of the $\Delta\Sigma$ modulator.

6.7.2 Analysis Results

The analysis tool of the previous section enables the evaluation and comparison of the effect of MASH and single-loop $\Delta\Sigma$ on the spectral purity of the fractional-N frequency synthesizer. As an example, an analysis is performed for a 3rd-order MASH modulator and the 3rd-order single-loop multi-bit $\Delta\Sigma$ modulator with an internal accuracy of 16 bit. The fractional division number is 67.92. With a reference frequency of 26 MHz, the output frequency is 1.76592 GHz, i.e. 2.08 MHz offset from an integer multiple of f_{ref}. At output frequencies close to an integer multiple of f_{ref}, the fractional-N synthesizer is more sensitive for spurious tones due to the worse randomization by the $\Delta\Sigma$ modulator. The analysis is performed with the following non-linearities: a 0.1% dead zone and a gain mismatch of $\pm 1\%$. The size of the introduced dead zone corresponds to $0.72°$ around $0°$ phase error. The gain mismatch is defined as the mismatch between the up and the down current of the charge pumps in $\pm$ percent.

In Fig. 6.19 (c) and (d), the time-domain phase error $\Delta\theta[i]$ is plotted for both modulators. Note that the $\Delta\Sigma$ fractional-N PLL frequency synthesizer can hardly be called a *phase-locked*

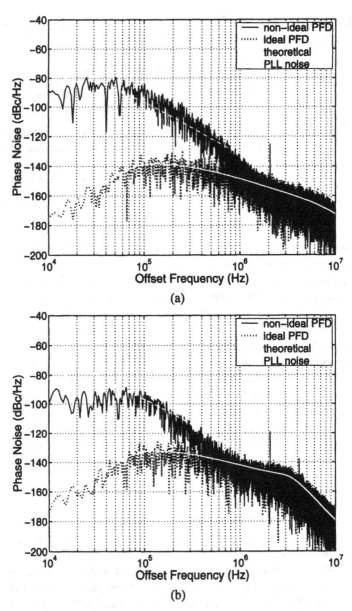

Figure 6.20: *Simulation results: The ΔΣ noise at the output of the PLL for (a) the MASH modulator and (b) the single-loop, multi-bit modulator. The results are plotted for an ideal PFD (dotted), which closely corresponds to the theoretical results (solid light grey) and for a non-linear PFD (solid). The results are compared to the simulated PLL phase noise without ΔΣ modulator (the dash-dotted line).*

oop, since the loop is never in lock! It is clear that the phase error due to the MASH modulation is much larger than that due to the single-loop modulation. To clarify this, the time-domain representation of the switching of the division modulus is revisited in Fig. 6.19 (a) and (b). The more intensive switching of the MASH modulator, which is reflected in the frequency domain by increased high frequency noise, is the cause of the large instantaneous phase error. Due to the shaping of the HF noise in the single-loop $\Delta\Sigma$ modulator, i.e. less modulus switching, the instantaneous phase error is seriously decreased in the single-loop $\Delta\Sigma$ modulator. This has two important consequences; First, the on-time of the charge pumps is smaller for the single-loop modulator, making it less sensitive to noise coupling from the substrate and the power supply. Secondly, the sensitivity to the non-linear $\Delta\theta \rightarrow I_{qp}$ conversion in terms of noise leakage is reduced.

To examine the effect of non-linearities in the frequency domain, the FFTs of the charge pump current pulses $CP[i]$ are computed and plotted in Fig. 6.19 (e) and (f). Both phase noise plots show the spectrum of the normalized charge pump output current. A noise floor appears in the output spectrum as well as spurious tones, although the $\Delta\Sigma$ output is perfectly randomized and dithered. Due to the non-linear mixing in the PFD-charge-pump, noise at $f_{ref}/2$ folds back to lower offset frequencies, similar to the effect of noise leakage due to a non-linear DAC in a multi-bit $\Delta\Sigma$ ADC. Since the noise at $f_{ref}/2$ is much lower for the single-loop $\Delta\Sigma$ modulator, the corresponding noise leakage due to non-linear mixing in the PFD is reduced; In the time domain, this effect corresponds to the smaller phase excursions. The difference in phase error magnitude between MASH and single-loop modulators is thus reflected in a lower noise floor, i.e. a 10 dB difference. Another effect is the appearance of previously unnoticed spurious tones in the output spectrum at $j \times K/2^k \cdot f_{ref}$ with $j = 1, 2, 3, \ldots$. The origin of the spurious tones is discussed in the next section.

Fig. 6.20 shows the $\Delta\Sigma$ noise of both modulators as it appears at the PLL output for an ideal (dotted) and a non-linear $\Delta\theta \rightarrow I_{qp}$ conversion (solid). The results of the ideal case closely match the theoretical results of Section 6.6 (solid light grey). The dash-dotted line represents the simulated output spectrum of the PLL without $\Delta\Sigma$ modulator, discussed in Chapter 5. Due to the noise leakage, the in-band noise of the synthesizer is seriously contaminated by the $\Delta\Sigma$ noise, posing a possible thread to the $\Delta\Phi_{rms}$ specification, especially in the case of the MASH converter. Also the phase noise specification at 600 kHz offset is endangered by noise leakage, again especially for the MASH converter. For the single-loop modulator, the $\Delta\Sigma$ phase noise at 3 MHz is critical, as was predicted by the theoretical analysis. In both cases, the spurious tones, observed at the output of the PFD are higher than the simulated PLL noise. Their power is however much lower than the -80 dBc specification of DCS-1800. The spurious tones are more critical when synthesizing frequencies closer to an integer value, as discussed in Appendix B.

An analysis sweep is performed, to study the influence of different degrees of non-linearity on the output phase noise of the synthesizer. To get an idea on the mechanisms behind noise leakage, the analysis is swept over different degrees of gain mismatch without dead zone and different sizes of the dead zone without gain mismatch. In Fig. 6.21, the results of this sweep are plotted for the 3rd-order MASH ((a) and (b)) and single-loop modulator ((c) and (d)). For the non-linearities, log-scale is applied to show the linear relationship of the noise leakage and the non-linearities.

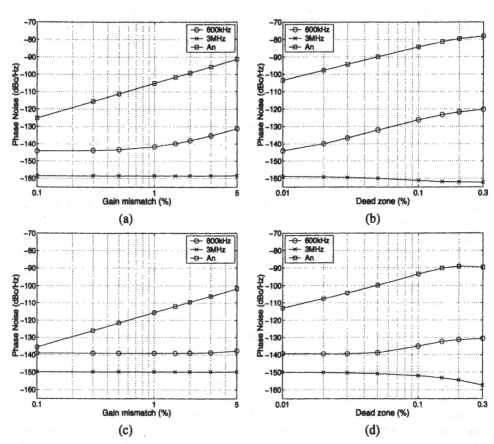

Figure 6.21: *Simulation results: The simulated output phase noise at 600 kHz, 3 MHz and in-band noise A_n, respectively, versus the gain mismatch in the $\Delta\theta \to I_{qp}$ conversion in $\pm\%$ and the dead zone in % of the PFD phase range for the MASH $\Delta\Sigma$ modulator ((a) and (b)) and the single-loop, multi-bit $\Delta\Sigma$ modulator ((c) and (d)).*

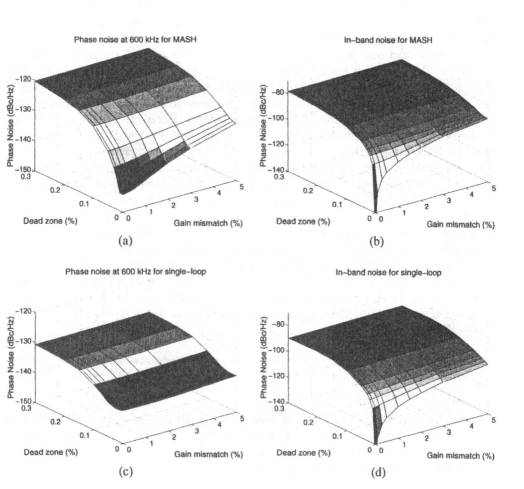

Figure 6.22: *Simulation results: The simulated output phase noise of the MASH $\Delta\Sigma$ modulator (a) at 600 kHz and (b) the in-band phase noise versus gain mismatch and dead zone. The simulated output phase noise of the single-loop $\Delta\Sigma$ modulator (c) at 600 kHz and (d) the in-band phase noise versus gain mismatch and dead zone.*

Gain mismatch in the charge pumps gives rise to a soft non-linearity in the $\Delta\theta \rightarrow I_{qp}$ conversion, and hence non-linear mixing of the $\Delta\Sigma$ noise. As can be seen in Fig. 6.21 (a) and (b), this results in an increased in-band noise A_n and in a higher phase noise at 600 kHz offset. The phase noise levels are however sufficiently low to satisfy DCS-1800 specifications, even for a gain mismatch of $\pm5\%$. For the single-loop $\Delta\Sigma$ modulator, the phase noise at 600 kHz is not influenced by the gain mismatch, due to the HF noise shaping.

However, a dead zone introduces a hard non-linearity in the $\Delta\theta \rightarrow I_{qp}$ conversion around the origin. This leads to a much higher noise leakage, even for very small dead zones (see Fig. 6.21 (b) and (d)). In the case of a MASH modulator, a dead zone of only 0.01% (or 0.07°), increases the in-band noise to almost -100 dBc. The phase noise at 600 kHz is increased in a similar way to levels that are critical for the phase noise specification. The problems are less severe for the single-loop modulator; The in-band noise remains below -90 dBc and the phase noise at 600 kHz is lower than -130 dBc/Hz. A second effect of the dead zone is that the loop gain is seriously lowered or even destroyed around 0°. As a consequence, the noise leakage for larger dead zones is slightly contained, which shows up in Fig. 6.21 (b) and (d) as the flattening of the increase of A_n and $\mathcal{L}\{600 \text{ kHz}\}$. The phase noise of the $\Delta\Sigma$ modulator at 3 MHz offset is even decreased by the lowering of the loop gain. The effect is more pronounced for the single-loop modulator, since the corresponding phase excursions are much smaller, enhancing the influence of the dead zone on the amplification of the noise to the synthesizer output.

In Fig. 6.22, the results of a simulation sweep of the combined effects of gain mismatch and a dead zone on the in-band noise and the phase noise at 600 kHz, is presented. Here, a linear scale is applied to illustrate the dramatic increase of the noise leakage for even small non-linearities compared to the ideal case. Some important conclusions can be drawn from the analysis results of both Fig. 6.21 and Fig. 6.22:

- Although the noise at 600 kHz is lower for a MASH modulator in the ideal case, only a small degree of non-linearity is enough to make the phase noise higher than for the single-loop modulator.

- The phase noise at 3 MHz is more critical for a single-loop modulator, due to the pole position shift in the noise transfer function. The noise at 3 MHz is virtually not influenced by gain mismatch. The decrease in phase noise at 3 MHz for increasing dead zone is due to the lowering of the loop gain around 0°.

- The in-band noise leakage in the case of a single-loop $\Delta\Sigma$ modulator is always lower than in the case of a MASH modulator. For the MASH modulator, noise leakage is highly critical in the presence of a dead zone.

- The influence of an even very small dead zone is much more severe than the influence of gain mismatch. The presence of a dead zone in the $\Delta\theta \rightarrow I_{qp}$ conversion must be avoided at all times.

For compactness and completeness, extra analysis results for fractional division by 67.9923, i.e. 200 kHz from an integer multiple of f_{ref} and for intermediate fractional division moduli are

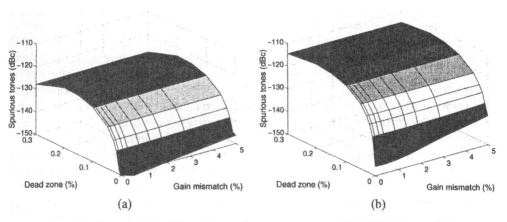

Figure 6.23: *Simulation results: The simulated spurious tones at 2.08 MHz for a division by 7.92 for the MASH (a) and the single-loop $\Delta\Sigma$ modulator (b) versus gain mismatch and dead zone.*

presented in Appendix B. The previous conclusions also apply to these simulation results and additional remarks are given.

6.7.3   Analysis Results: The Origin of Spurious Tones

The previous section focused on the influence of non-linearities on the output phase noise of the synthesizer. But what about the spurious tones that are visible in Fig. 6.20? When looking at the output spectrum of the $\Delta\Sigma$ modulators (Fig. 6.9 (b), Fig. 6.12 (b)), no spurious tones can be distinguished. But when looking at Fig. 6.19 (e) and (f) and Fig. 6.20, the previously unnoticed spurs appear at the output of the charge pump and hence at the synthesizer output. As discussed in Section 6.4.4, first-order $\Delta\Sigma$ modulators with DC inputs suffer from spurious tones in their output spectrum. The power and frequency of the spurious tones depend on the DC-input of the $\Delta\Sigma$ modulator (see Fig. A.2 in Appendix B). As stated in Eq. (A.2), the frequency of the spurious tones is the input fraction times the reference frequency and its harmonics. Several improvements to the first-order modulator are suggested to break the patterns and convert the spurious energy to white noise: moving to higher-order modulators, addition of a dither signal and multi-bit quantization.

The modulators in the analysis are both 3rd-order and have multi-level quantization. The dithering applied in the simulations originates from the uniformly distributed pseudo-random generator in MATLAB. This generator can generate all the floating point numbers in the closed interval $[2^{-53}, 1 - 2^{-53}]$. Theoretically, it can generate over 2^{1492} values before repeating itself. The dithering sequence is applied at the first integrator, together with the input signal. To make sure that the dithering does not create a noise floor in the $\Delta\Sigma$ output spectrum, the dithering at the input is n-th-order shaped, with n the order of the modulator. Dithering can be applied to the other integrators or at the input of the quantizer, with different orders of shaping. Here, a

1 LSB, 3rd-order shaped dither sequence is applied at the input of the first integrator. The study of the different kinds of pseudo-random generators and their influence on the performance of a fractional-N synthesizer is beyond the scope of this book.

The same analysis sweep as in the previous section is performed to investigate the influence of different degrees of non-linearity on the re-emerging of spurious tones in the synthesizer output spectrum. The results are plotted in Fig. 6.23. Although all three options to randomize the $\Delta\Sigma$ output to white noise are implemented, spurious tones still emerge after non-linear signal processing. The effect of gain mismatch in the charge pumps is much smaller than the effect of a small dead zone in the PFD. The MASH modulator suffers less from spurious tones, since it provides better randomization of its output by using almost all possible output codes to obtain the wanted mean output. However, the synthesizer's sensitivity to noise coupling through the substrate and supply lines is increased by the MASH modulator, due to the longer on-time of the charge pumps . The single-loop modulator is less sensitive but its output is inherently less randomized, resulting in higher and even second-order spurious tones at the PLL output (see Fig. 6.20). The power of the spurious tones remains below -115 dBc and is consequently not critical for the DCS-1800 system for a division by 67.92. For large dead zones, the spur level decreases due to the decrease in loop gain. The effect is more pronounced for single-loop modulators. In Appendix B, a similar analysis sweep is presented for fractional division by number in between and closer to integer moduli, i.e. 67.577 and 67.9923. In the first case, spurious tones are well below -150 dBc, meaning that the tones are below the phase noise and pose no danger to the system. In the second case however, even small degrees of non-linear behavior result in spurious tones higher than -80 dBc, especially for the single-loop modulator. In practice, fractional division by numbers close to integer moduli must be avoided or the $\Delta\theta \rightarrow I_{qp}$ conversion must be made highly linear to enable the use of $\Delta\Sigma$ fractional-N synthesizers in high quality telecommunication systems.

The emerging of spurious tones can be explained by the fact that the $\Delta\Sigma$ modulator is unable to sufficiently decorrelate the successive output samples. To quantify the correlation in the $\Delta\Sigma$ modulator output, the time variation of the output sequence is described using the discrete time autocorrelation estimate, $R_{yy}(m)$ [Nors97].

$$R_{yy}(m) = \frac{1}{2N} \sum_{j=-N}^{N-1} y(j)y(j+m) \tag{6.38}$$

The discrete-time autocorrelation estimate is plotted for both modulators for inputs close to an integer value, i.e. division by 67.92 for 4000 shifts (see Fig. 6.24). If the output sequence is perfect white noise, $R_{yy} \rightarrow 0$ for $m \neq 0$. The autocorrelation estimate calculations show correlation in the $\Delta\Sigma$ output, although 1 LSB noise shaped dithering is applied. The autocorrelation of the single-loop $\Delta\Sigma$ modulator shows larger correlation peaks, explaining the higher spurious tones in the output phase noise spectrum of the PLL. The necessary internal accuracy of the $\Delta\Sigma$ modulators can be determined such that the autocorrelation estimate of the modulators approaches that of white noise. The minimal internal accuracy is found to be at least 13 bits for MASH and 18 bit for single-loop modulators to sufficiently decorrelate the $\Delta\Sigma$ modulator output for inputs close to integers. Unfortunately, no direct relation is found between the outlook of

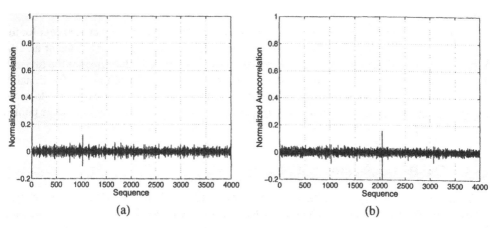

Figure 6.24: *The discrete time autocorrelation estimate of the modulator outputs for (a) the MASH modulator and (b) the single-loop, multi-bit modulator with 16 bit internal accuracy and for a fractional division of 67.92.*

the autocorrelation estimate and the power of the spurious tones at the output of the synthesizer. Even for high internal accuracies, where the autocorrelation estimate is close to that of white noise, tones are present at the synthesizer output.

A second possible source of tones is the down-conversion of tones which are inherently present around $f_{ref}/2$ [Nors97], by the non-linear mixing in the PFD. This effect can be worsened by substrate and power supply coupling with signals at $f_{ref}/2$. Note that all simulations are performed without taking into account noise coupling through the substrate or power supply lines. As a consequence, the actual spurious performance of the $\Delta\Sigma$ fractional-N PLL is likely to be worse than simulated, if no special precautions are taken to shield the $\Delta\theta \rightarrow I_{qp}$ conversion from incoming noise.

6.8 The Fractional-N Synthesizer Circuit Design

The goal of this book is the integration in standard CMOS technology of a $\Delta\Sigma$ fractional-N frequency synthesizer to enable the full integration of a transceiver front-end in CMOS, including a low-IF receiver and a direct upconversion transmitter [Stey00b], which is discussed in Chapter 7. To prevent degradation of the spectral purity by digital noise coupling, the $\Delta\Sigma$ modulator is scheduled for integration on the digital base-band signal processing IC of the full transceiver system. Therefore, to implement the $\Delta\Sigma$ fractional-N synthesizer, the fully integrated type-II, 4th-order phase-locked loop of Chapter 5 is chosen. To check the validity of the previous analysis for different types of $\Delta\Sigma$ modulators, the $\Delta\Sigma$ modulator is not integrated on the same die as the PLL[1]. For measurements, the prescaler moduli are controlled by an externally generated $\Delta\Sigma$

---

[1] $\Delta\Sigma$ modulator integration was scheduled for the next version of the synthesizer IC, but access to the licensed Toshiba 0.25μm CMOS technology was denied. Therefore, a full redesign would have been performed in a different

bitstream.

The design of the building block circuits of the PLL is elaborated in Chapter 5, except for the building blocks responsible for the $\Delta\theta \to I_{qp}$ conversion: the phase-frequency detector and the charge pumps. The design of the PFD and charge pumps is left for this chapter since the linearity of those building blocks is of utmost importance for the spectral purity of the $\Delta\Sigma$ fractional-N synthesizer. First, the PFD design is discussed with the focus on the dead zone and how to avoid it. Secondly, the mechanisms of spurious tone generation and gain mismatch in the charge pumps are investigated.

6.8.1 The Phase-Frequency Detector

In Chapter 2, different types of phase detectors are summarized. In the $\Delta\Sigma$ fractional-N synthesizer, a dead-zone-free phase-frequency detector is realized (see Fig. 6.25), by implementing a reset delay [Gard79, Miju94]. The most important problem of the PFD is the existence of a dead zone around zero degrees phase errors; When the loop is in lock and a small phase error occurs, the PFD needs a finite period of time to respond and register the change on the $\overline{Up}$ and $\overline{Dn}$ outputs. The time period depends on the propagation delays within the logic gates of the device. All this time the output of the charge pumps, which is controlled by the PFD outputs, remains in its high impedance state and the PLL is not locked. Therefore, the VCO phase noise is transferred to the output in a bandwidth, corresponding to the size of the dead zone, resulting in the spreading out of the peak in the PLL output spectrum. A second consequence of the dead zone is that it introduces a hard non-linearity in the $\Delta\theta \to I_{qp}$ conversion, giving rise to noise leakage and spurious tones in the case of a $\Delta\Sigma$ fractional-N synthesizer. The size of the dead zone can be quite large, especially with high reference frequencies [Hill92]. The dead zone is simulated in HSpice using the actual PFD circuit of Fig. 6.25 with a delay of 0 ns and the result is presented in Fig. 6.26 (a). A dead zone of 0.8% of the PFD range is clearly noticeable. What's more, up to 5% of the available phase error range is highly non-linear, which is the same range over which the phase error, generated by a $\Delta\Sigma$ modulator control, varies. In other words, the classical PFD circuit is not fit for use in $\Delta\Sigma$ fractional-N synthesizers.

The delay block after the 4-input NAND in Fig. 6.25 gives a fixed minimum width to the output pulses, determined by the size of the delay, T_d. If the divider output lags the reference signal, the Up signal becomes one for a time period, determined by the time difference between the two inputs and the delay time of the circuit, $T_c + T_d$, with T_c the circuit delay in the classic PFD. Also the Dn output becomes one for a time period, determined by the delay in the circuit, $T_c + T_d$ (see Fig. 6.25). The net difference between the Up and Dn outputs controls the amount of current the charge pump delivers to the filter, which is proportional to the phase difference between the two inputs. When the loop is in lock, both Up and Dn outputs are one for an equal time period. As a result, both the up- and down-current source are on, such that the net current injected in the filter impedance is zero, but the charge pump output is never in its high impedance state. In other words, the loop is locked even for small phase difference and the dead zone is vanquished. The delay in the PFD circuit is implemented by two inverters with extra

technology, which was impossible within the given time frame.

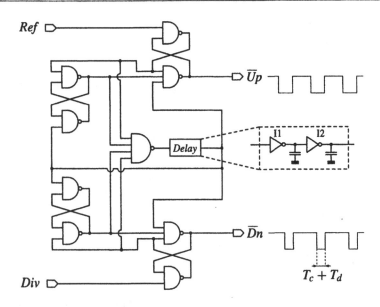

Figure 6.25: *The dead-zone-free phase-frequency detector. The two inverters introduce the necessary delay to enhance the PFD sensitivity for small phase errors.*

capacitors added to control the delay. The capacitors are both 130 fF, resulting in a delay of 2.85 ns. This means that even when the loop is locked the charge pumps are on for 7.5% of the time. The dead zone is again simulated for the circuit of Fig. 6.25, resulting in the $\Delta\theta \rightarrow I_{qp}$ conversion characteristic plotted in Fig. 6.26 (b). The simulated conversion is perfectly linear and no dead zone is present. The choice of the delay is a trade-off between overcoming the dead zone and sensitivity to noise coupling. The delay, i.e. the on-time of the charge pumps must be made as small as possible, since at that moment the pumps are sensitive to noise coupling through the substrate or the power supply lines. A higher on-time also leads to higher charge pump current noise, which can corrupt the in-band noise of the PLL.

Some other solutions exist to get rid of a dead zone. First, the dead zone region could be moved up or down the VCO's input range by injecting offset voltage at the input of the VCO. The dead zone is avoided, but the overall phase detection range of the PFD is reduced, which is defeats one of the original advantages of the digital PFD architecture. A second option is the phase-frequency discriminator [Hill92]. At lock the output of the discriminator is a square wave with an exact 50% duty cycle. When a phase error occurs between the two inputs, the duty cycle of the square wave is changed. By filtering, the DC of the square wave is extracted and used as a measure for the phase error. It is clear that the phase-frequency discriminator can react to small phase errors, so it doesn't suffer from a dead zone. But, since the output is a square wave, the spurious suppression is worse than that of the dead-zone-free PFD.

The addition of a delay in the critical path of the PFD also influences the maximal input frequency for phase and frequency detection. Since the increase in reference frequency is the

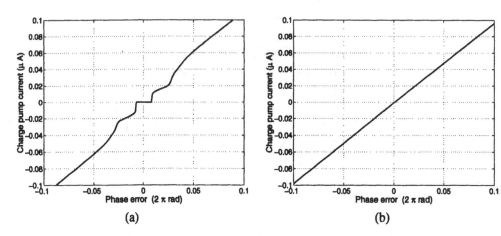

(a)   **(b)**

Figure 6.26: *The* $\Delta\theta \to I_{qp}$ *conversion characteristic of a phase-frequency detector with (a) no delay and (b)* 3 ns *delay.*

major advantage of fractional-N synthesis, the frequency capabilities of the PFD are of major importance. In [Soy90], the frequency limitations of a conventional PFD, i.e. without delay are discussed; If the VCO output lags the reference signal, *Up* is set high. A following transition in the VCO output sets *Dn* high, creating a reset pulse at the output of the 4-NAND, which resets both outputs. Due to the finite response time of the digital circuit, the reset pulse has a finite width equal to the delay in the circuit, $\Delta R = T_c$. This means that a time ΔR is needed to start the phase detection of the next inputs. When the input frequency to the PFD, f_{ref}, is equal to $1/(2\Delta R)$, the $\Delta\theta \to I_{qp}$ characteristic becomes a sawtooth; the PFD is no longer capable of frequency detection and the useful phase range is reduced to $\{-\pi, \pi\}$. Beyond this frequency, the PFD outputs wrong information. In this design case, ΔR is approximately 0.8 ns, setting the maximum for f_{ref} to 625 MHz. However, to get rid of the dead zone, a delay of 2.85 ns is added in the critical path and ΔR becomes $(T_c + T_d) \approx 3.65$ns. The maximum input frequency is reduced to 274 MHz. This is of course much higher than the proposed reference frequency of 26 MHz, but high frequency effects are already visible in the $\Delta\theta \to I_{qp}$ conversion. In Fig. 6.27, the $\Delta\theta \to I_{qp}$ conversion is plotted for a f_{ref} of 26 MHz for the designed PFD. The PFD will make wrong decisions at around -2π rad phase error, resulting in slower settling of the PLL. Also, the phase detection range is decreased, but the phase error range due to the $\Delta\Sigma$ modulator control is not harmed. The $\Delta\theta \to I_{qp}$ conversion in the case of a 3 ns delay is also plotted (only for positive phase errors). The high frequency effects are worsened by the delay.

6.8.2   The Charge Pumps

The second source of non-linearity in the $\Delta\theta \to I_{qp}$ conversion is the gain mismatch in the up- and down-currents of the charge pumps. Basically, the charge pumps are current sources which are switched on and off by the PFD output to deliver the right amount of charge to the loop filter to maintain loop lock. Since the switching occurs at every down flank of the reference frequency,

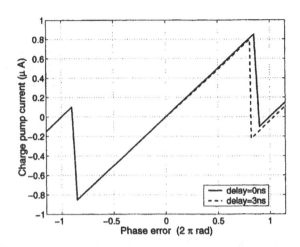

Figure 6.27: *The total* $\Delta\theta \rightarrow I_{qp}$ *conversion characteristic of the PFD with (a) zero delay and (b) with a delay of* 3ns.

charge injections at the switching times modulate the VCO frequency, such that spurious tones appear at an offset of f_{ref} from the carrier. In the presented design, two charge pumps are integrated for the dual-path loop filter (see Chapter 5): one unity charge pump and a charge pump which delivers a B times higher current, the B-charge pump. The design of the charge pumps comes down to maximizing the spurious suppression and minimizing the gain mismatch.

6.8.2.1   The Spurious Suppression

A conventional charge pump PLL does not suffer from unwanted charge injection in the loop filter when the loop is locked. In lock, both up- and down current sources are off. By introducing the zero-dead-zone PFD, the additional delay in the PFD reset path turns both current sources on, even in lock, for an equal amount of time. Ideally, no charge is injected in the loop filter. However, at the switching moments unwanted charge is injected, leading to high spurious tones at the reference frequency in the synthesizer output spectrum. To remedy this, several precautions are taken:

First, the current sources themselves are never switched on or off to prevent current switching effects on the drains of the current sources. Instead, when the charge pump is in its off-state, the current is redirected to a dummy current branch. In Fig. 6.28 (a), the charge pump circuit is shown. The dummy current branch is at the left and its output is connected to the reference voltage, which is $V_{dd}/2$. The control signals for the dummy branch are denoted by an end-suffix d and are the inverse of the control signals of the output branch. Only when the output branch is closed all current goes through the dummy branch. Since the current sources always conduct current, no start up delays occur in the circuit and the charge pump immediately responds to changing control signals.

Secondly, charge injection is carefully minimized for suppression of reference spurious tones.

There are two effects that determine the charge injection when the MOS transistors turn off; First, channel charge present in the channel when the MOS is on, must flow out of the channel region to the source and drain junctions. Since the V_{DS} of the switch transistors in Fig. 6.28 is zero, the channel charge is given by Eq. (6.39). The second charge injection comes from the overlap capacitances between the gate and the junctions. Both effects result in a voltage change at the output of the charge pump; If the control signals of the MOS switches change sufficiently fast, the channel charge flows equally out through both junctions [John97].

$$Q_{ch} = WLC_{ox}(V_{GS} - V_T) \tag{6.39}$$

$$\Delta V_{ch} = \frac{Q_{ch}/2}{C_{out}} \tag{6.40}$$

$$\Delta V_{ov} = \frac{V_{dd}/2 \cdot W C_{ov}}{W C_{ov} + C_{out}} \tag{6.41}$$

Therefore, latches are placed at the output of the timing control circuit of Fig. 6.28 (b). Due to the fast switching, only half of the channel charge flows in the low impedance at the output of the charge pump. The resulting voltage change ΔV_{ch} at the charge pump output is given in Eq. (6.40) with C_{out} the output capacitance of the charge pump, i.e. C_z for the unity charge pump and C_p for the B-charge pump. The voltage change due to the overlap capacitance, ΔV_{ov}, is given in Eq. (6.41) and comes from the capacitive division of the voltage over the overlap capacitance between the C_{ov} and C_{out}.

By implementing the NMOS switches of the charge pump with minimal sizes, the parasitic charge injection is an order of magnitude smaller than the useful charge injected by the charge pump. In parallel with the NMOS switches, PMOS switches are placed to minimize the charge injection due to the overlap capacitances. The size of the PMOS transistors is such that the injected charge is equal but with opposite sign to the NMOS overlap charge. The overlap capacitance of the PMOS is 75 % larger than that of the NMOS, such that the NMOS width is made 75 % larger than that of the PMOS, i.e. 3.5 μm. By implementing the switches with N- and PMOS transistors, the overlap charge injection is minimized but the channel charge injected is theoretically doubled. However, the bulks of the NMOS switches are connected to the ground and those of the PMOS to V_{dd}. Due to the bulk effect, the threshold voltages, V_T, for N- and PMOS become rather high, 0.8V and 0.95V, respectively. As a result, the PMOS switches are hardly on and their channel charge is more than an order of magnitude smaller than that of the NMOS switches. The bulk effect also reduces the channel charge of the NMOS switches to the level of the overlap charge injection. So ideally, the PMOS switches cancel the overlap charge injection of the NMOS switches but only contribute marginally to the channel charge injection, such that unwanted charge injection is restricted to only half of the channel charge of the NMOS switches.

To realize the control signals for the N- and PMOS switches, a customized timing scheme is implemented. The names of the control signals are chosen in a systematic way; The first letter of the suffix is u or d, for up and down, denoting how the switches influence the frequency of the VCO. In the design case, the u-switches pull current out of the loop filter, but due to an inversion in the loop filter adder, the control voltage goes up, increasing the VCO frequency.

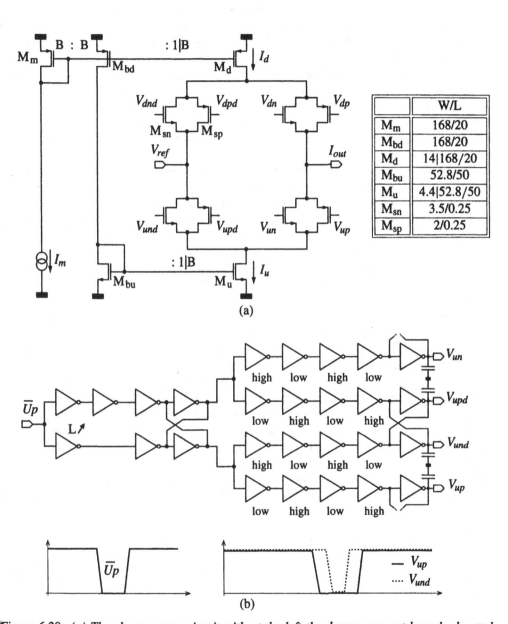

Figure 6.28: *(a) The charge pump circuit with at the left the dummy current branch, denoted by the suffix d and the output branch. (b) The timing control circuit and signals to control the dummy and the output current branch switches of the charge pump.*

The second letter of the suffix is n or p, for NMOS or PMOS control respectively. To minimize charge injection, the control signals for the N and PMOS switches must be perfectly differential and have to be switched sufficiently fast. As an example the timing control circuit is drawn in Fig. 6.28 (b) for the Up signal path. To realize the 180° phase shift, a simple inverter is not sufficient. One control signal always lags the other, such that the overlap charge injection is insufficiently compensated. Therefore, both control signals go through an inverter string, one with two inverters and one with one inverter, with the same global delay. The delay of the single inverter is increased by enlarging the gate length L of both transistors. To ensure the 180° and to equalize fall and rise times of both signal paths, a minimal sized latch is implemented at the end of the inverter strings.

The circuit, succeeding to the differential switch control circuit in Fig. 6.28 (b), controls the timing of the switches in the dummy branch with respect to the switches of the output branch. In theory, pure differential control of respective branches would do, such that one branch closes when the other opens. However, due to the finite switching time, it can occur that both current branches are off at the same time for a short period of time. This situation must be avoided at all time. Therefore, the timing control is modified such that the dummy branch closes *after* the output branch is opened and the dummy branch opens *before* the output branch is closed. In that way, the current sources always conduct current. This is accomplished by modifying the threshold voltages of the successive inverters in the second inverter string. As an example, the generation of the control voltage of the PMOS up-switch in the output branch, V_{up} and V_{und} is elaborated. If the $\overline{Up}$ signal of the PFD is as in Fig. 6.28 (b), the signal at the input of the second inverter string is an inverted copy of $\overline{Up}$. When this signal passes first through an inverter of which the threshold is lowered (bottom inverter string), the NMOS of the inverter reacts faster than the NMOS of the inverter with a higher threshold. The output of the "low" inverter goes to its logic "0" faster than the "high" inverter. When going to its logic "1" state, it is slower than the "high" inverter. As a result, the input pulse in the bottom inverter string is broadened towards the output V_{up}, while the output pulse V_{und} of the other inverter string is narrowed, as depicted in Fig. 6.28 (b). The following inverters have the same effect on the signal pulse. In this way, the wanted dummy branch timing scheme is achieved. To further optimize the spurious suppression, latches are placed at the outputs of the second inverter strings, to ensure differential control for N and PMOS switches and to equalize the rise and fall times of the control signals for optimal charge compensation. The latches also ensure sufficiently fast switching to minimize channel charge injection.

Last but not least, to minimize the spurious suppression the loop filter is implemented partly as an active filter(see Section 5.3.1). The active integrator at the output of the unity charge pump holds the charge pump output at V_{ref}. At the output of the B-charge pump is a passive RC-filter is implemented, but the resistor R_p provides a DC-path to V_{ref}, fixing the output voltage.

6.8.2.2 The Gain Mismatch

The other important issue in the design of charge pumps for fractional-N frequency synthesizers is the minimization of the gain mismatch in the $\Delta\theta \rightarrow I_{qp}$ conversion.

The measures taken to maximize the spurious suppression positively affect the gain mismatch; First, due to the active implementation of the loop filter, the current source transistors are always in saturation, independent of the VCO control voltage. This means that if the charge pumps are properly designed, the up- and down-currents, I_u and I_d, are equal and remain equal. In case of gain mismatch, an unwanted amount of charge is injected in the loop filter every reference cycle. The loop settles to a small phase offset that compensates the current mismatch. The net charge injected in the filter is zero, but reference spurs will appear at the synthesizer output. Secondly, the timing control scheme of Fig. 6.28 (b) ensures that the currents delivered to the loop filter are those of the current sources by avoiding current source start up delays and switching effects.

The design for minimal gain mismatch now comes down to accurately mirror the current I_m of the mother current source to the up- and down current sources. The mother current I_m is 15.6μA such that the current in the unity charge pump is equal to 1.3μA with a multiplication factor B of 12. Fig. 6.28 (a) shows how the mirroring is performed in both charge pumps. The current from the mother source is mirrored to a common current branch for up- and down-currents, forcing them to be equal. From this branch the current is mirrored to the two charge pumps, with their respective mirror factors. The $V_{GS} - V_T$ of the current sources is chosen as a trade-off between low noise and low-voltage operations. A third factor determining this trade-off is the current mirror accuracy. To rule out random current variations in the current mirroring due to variations in the threshold voltages, the $V_{GS} - V_T$ must be chosen as high as possible. The trade-off $V_{GS} - V_T$ value is 0.3V. There is also a systematic current variation due to the difference in V_{DS} of the different transistors. This is remedied by choosing large gate lengths for the current source transistors. The sizes of the different transistors are listed in Fig. 6.28 (a). The systematic error can be reduced by implementing a cascode current mirror, which is not done here to enable low-voltage operation. Also the matching of the transistors themselves is important, since their actual sizes determine the accuracy of the current mirror factor $B : B$ or $B : 1$. The size of the current in the mirroring branch is taken B such that the W/L of the corresponding transistors is larger, which is better for matching. Since the W/L of all the current transistors is low due to the large $V_{GS} - V_T$ and the small current, the gate length of the transistors is again chosen large to improve the transistor matching. A secondary effect is that the 1/f noise of the charge pumps is lowered.

This current mirror topology is capable of sub-2V operation. The calculated worst-case gain mismatch is below $\pm$0.5%.

6.9 Experimental Results

6.9.1 Measurement Setup

To test the functionality of the presented $\Delta\Sigma$ fractional-N synthesizer, a $\Delta\Sigma$ bit stream is externally generated and fed to the modulus control of the synthesizer IC. Fig. 6.29 shows a microphotograph of the IC and the measurement setup in which it is embedded. The reference frequency is generated by a 26 MHz crystal oscillator to ensure that the in-band noise of the synthesizer is not corrupted by the reference signal. The fractional measurements are performed by controlling

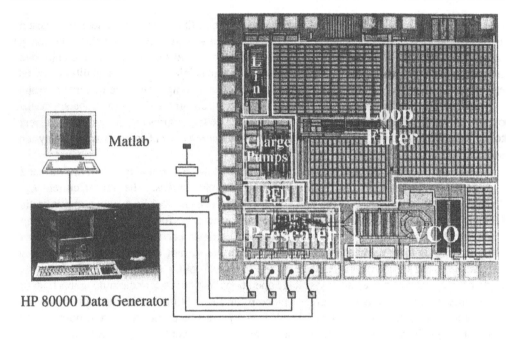

Figure 6.29: *The IC microphotograph and the measurement setup in which it is embedded.*

the frequency divider moduli with a HP80000 data generator. The data generator outputs a 4-bit, 2V rail-to-rail control word that is fed to the integrated synthesizer IC. The 4-bit ΔΣ output bit stream is generated with transient simulations of ΔΣ modulators in Matlab. In this way, a flexible measurement setup is realized that enables the investigation of different types of ΔΣ modulators and their influence on the performance of a ΔΣ fractional-N synthesizer. Moreover, different clocking schemes can be studied, by controlling the delay between the reference clock of the synthesizer and the clock of the ΔΣ modulator.

Apart from providing flexibility, the measurement setup also poses some problems. First, the clocks of the modulator and the synthesizer must be synchronized to make sure that the ΔΣ modulator provides its frequency control at the right time. In this measurement setup, the reference crystal oscillator signal is used as the mother clock for the total system. The data output of the HP data generator is clocked externally by the mother clock.

A second problem is the switching power supply of the data generator. The switching frequency is 25 kHz with a high harmonic content at 50 and 75 kHz. A lot of effort has been put in shielding the IC from these high unwanted signals, which appear in the synthesizer output spectrum as spurious tones. The copper measurement box has been put inside a grounded steel box with external connectors. The grounds of these connectors is connected to the steel box, shielding the ground of the data generator from the ground of the IC. In this way, the coupling of the unwanted power supply signals could be reduced to negligible levels.

A third problem is the high jitter of the data generator output signals, which is 10 ps typical

md < 25 ps maximal [HP93]. This directly shows up in the overall synthesizer spectrum as
m increase of the in-band noise floor, which is simulated (Maltab) to be more than -70 dBc/Hz
around 10 kHz offset due to jitter of the data generator. The problem is even worse, since the data
generator determines the level of its output signals in function of how they are terminated. In the
measurement setup, the output of the data generator is connected to the measurement PCB by
a BNC connector. This connector is connected to the ceramic measurement substrate on which
the IC is glued. The connection between the substrate and the IC is a bond wire. As a result, the
termination is not well defined and reflections occur, such that the control signal, reaching the IC
is a only square-wave-like with an undetermined low and high level and lots of ringing, leading
again to increased in-band noise in the synthesizer.

Together with the high jitter, the most stringent problem comes from the limited memory capacity of the data generator, which limits the minimum measurable offset frequency. To measure
the fractional performance of the frequency synthesizer, Matlab data is stored in the data generator memory. Unfortunately, the maximum memory capacity of the data generator is only 128 kb.
To measure the phase noise of the synthesizer, the data in the memory is put in a loop to get a
continuous $\Delta\Sigma$ bit stream. Due to the repetition, large spurious tones show up in the synthesizer
spectrum at low offset frequencies, i.e. $f_{ref}/2^{17} \approx 200$ Hz and its harmonics. The phase noise of
the synthesizer is measured with a dedicated phase noise measurement system, the Europtest PN
9000. Since this system has a limited dynamic measuring range, gain calibration is performed every offset frequency decade. Due to the large tones at low offset frequencies, the gain calibration
for the corresponding offset frequency decades is erroneous, such that meaningful measurements
of the phase noise at offsets smaller than 10 kHz are not feasible. Also measurements in the offset
decade between 10 and 100 kHz are affected by the limited data generator memory[2].

6.9.2   Measurement Results

All presented measurements are performed with a 26 MHz reference frequency for a fractional
division by 67.92 for comparison with the simulated results. This fractional frequency is chosen because it is close to an integer multiple of the reference frequency. The input to the $\Delta\Sigma$
modulators is a 16-bit word (k=16), resulting in a frequency resolution of around 400 Hz. The
power supply voltage is only 2V. In Table 6.1, the loop properties and the simulated phase noise
contributions of all building blocks are summarized for the fractional operation at 1.76592 GHz.
Note that the $\Delta\Sigma$ modulator noise seriously corrupts the phase noise at 600 kHz.

Fig. 6.30 shows the output spectrum of the fractional-N PLL over a span of 55 MHz. The
reference spurs are well below -75 dBc, due to the careful charge pump timing control. Compared to the measured results of the integer-N PLL of the previous chapter, the reference spurs
are above the noise floor of the measurement setup. The worse reference spur suppression is
mainly due to the lower frequency at which is measured; The K_{vco} and thus the open-loop gain

---

[2]The integration of the digital $\Delta\Sigma$ modulator is not a highly complex task and was intended for the next version
of the DCS-1800 frequency synthesizer. The final integration step could not be accomplished since access to the
0.25μm technology, which was licensed by Toshiba, was denied. A full redesign in a different technology was
impossible within the given time frame

Loop Parameters		Loop Passives		Performance (at 600 kHz)	
ω_c	35 kHz	R_p	3.2 kΩ	$\mathcal{L}_{qp,z}$	-166.4 dBc/Hz
I_{qp}	2 μA	R_4	1.07 kΩ	$\mathcal{L}_{qp,p}$	-150.5 dBc/Hz
B	12	C_p	240 pF	$\mathcal{L}_{add}$	-133.0 dBc/Hz
		C_4	710 pF	$\mathcal{L}_{int}$	-137.0 dBc/Hz
		C_z	450 pF	$\mathcal{L}_{R_p}$	-134.6 dBc/Hz
		C_{tot}	1.4 nF	$\mathcal{L}_{R_4}$	-132.5 dBc/Hz
				$\mathcal{L}_{tot,int}$	-124.3 dBc/Hz
				$\mathcal{L}_{MASH}$	-123.6 dBc/Hz
				$\mathcal{L}_{SL}$	-128.0 dBc/Hz
PM	55°	L	2.857 nH	$\mathcal{L}_{tot,MASH}$	**-120.9 dBc/Hz**
ζ	0.72	Q_L	9	$\mathcal{L}_{tot,SL}$	**-122.8 dBc/Hz**

Table 6.1: *Summary of the loop properties and simulated performance of the ΔΣ fractional-N synthesizer at 1.76592 GHz for both modulators*

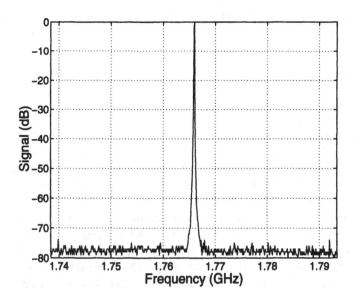

Figure 6.30: *The measured output spectrum of the ΔΣ fractional-N PLL at 1.76592 GHz. All spurious tones are well below −75 dBc/Hz.*

is approximately 30% higher in this operating point, resulting in the observed degradation of the spurious suppression.

The measured phase noise of the PLL with the MASH modulator and the single-loop, multi-bit modulator is presented in Fig. 6.31 and Fig. 6.33. Small spurs are present at 2.08 MHz as predicted by the simulations in Fig. 6.20. The overall fractional spur level is well below -100 dBc. The phase noise at 600 kHz is lower than -120 dBc/Hz for both modulators.

In Fig. 6.31, the measured phase noise of the PLL with the multi-bit, single-loop modulator (dark) is compared to the phase noise (a) of the PLL at integer division (light) by 68 and (b) of the stand-alone VCO (light). The phase noise of the synthesizer at integer division is as low as -124 dBc/Hz at 600 kHz offset, which is only 0.3 dB higher than predicted by the PLL simulations (see $\mathcal{L}_{tot,int}$ in Table 6.1). The shape of the phase noise differs from the measured phase noise spectrum in the previous chapter and the noise at 600 kHz is 1.5 dB higher due to the difference in operating frequency. The overall phase noise of the PLL with $\Delta\Sigma$ control is higher than without. Fig. 6.31 (b) shows that although the VCO has a high $1/f^3$ noise corner, the in-band noise is mainly determined by the PLL building blocks, the limited memory and high jitter of the data generator.

The higher noise at intermediate offset frequencies (between 50 kHz and 1 MHz) is predicted by the non-linear analysis. However, higher noise at low and high offset frequencies seriously deteriorates the spectral purity of the synthesizer. The noise at lower offsets originates only for a small part from noise leakage in the PFD. The rms phase error $\Delta\Phi_{rms}$ is increased from 1.7° to more than 3°. The measured results for fractional division are also much noisier than that of the integer measurement. The phase noise increase is due to the limited memory and the high jitter of the data generator (which can be proven in simulation). The measured phase noise in the frequency decade below 10 kHz is simply useless. To enhance the measurement accuracy, the memory of the data generator should be upgraded or the $\Delta\Sigma$ modulator should be integrated on the die. The noise at higher offset frequencies is corrupted by noise coupling from the data generator. As can be seen in Fig. 6.29, the $\Delta\Sigma$-control bonding wires, which conduct large and noisy control pulses (due to the high jitter and unknown signal termination) are close to the LC-tank and the bonding wires of the VCO power supply. Without proper shielding, the VCO phase noise is degraded by this noise coupling.

In Fig. 6.33, the measured phase noise of the synthesizer for (a) the MASH modulator and (b) the single-loop modulator and the $\Delta\Sigma$ noise as simulated in Section 6.7 (dashed) is compared. The dashed-dotted line is the simulated phase noise of the PLL without $\Delta\Sigma$-control. The simulated $\Delta\Sigma$ noise leakage closely matches the measured results. The extra delay in the PFD increases the on-time of the charge-pumps, making them more sensitive to the noise coupling from the data generator. The noise from the data generator affects the power supply on the die by inductive coupling between the bonding wires, leading to more in-band noise and overall noisier measurements. At low offsets the measured results deviate from the simulations due to the limited memory and high jitter of the data generator. Tones at around 500 kHz are not predicted by simulation, but are believed to come from sub-harmonic tones present in the $\Delta\Sigma$ modulator output [Nors97].

When comparing the results for the MASH and the single-loop modulator, the measured results are less pronounced than the simulated results (see Fig. 6.20). The measured phase noise

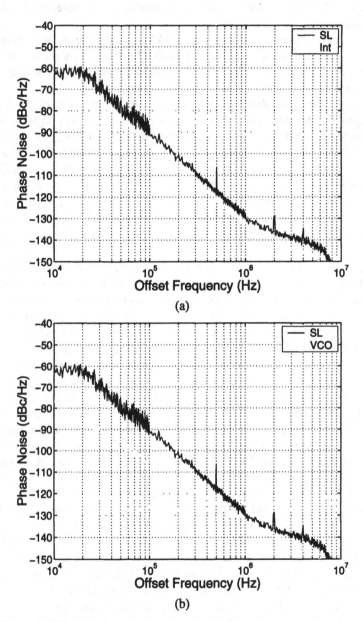

Figure 6.31: *Phase noise measurement (a) with the $\Delta\Sigma$ single-loop, multi-bit converter at 1.76592 GHz compared to the phase noise at integer division (light) and (b) compared to the phase noise of the stand-alone VCO.*

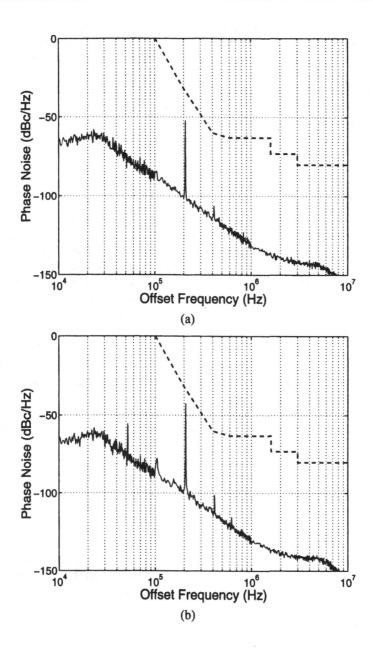

Figure 6.32: *Phase noise measurement with the (a) MASH and (b) the single-loop $\Delta\Sigma$ modulator for a division by 67.9923, which is the worst-case situation in terms of spurious tone suppression. Measurements comply to the DCS-1800 spurious suppression mask (dashed).*

for the single-loop modulator is however a few dBs lower than for the MASH modulator at intermediate offset frequencies. The measured in-band noise is the same for both modulators, although both the theoretical and the non-linear analysis predict a 6 to 10 dB difference. The overall in-band noise in both measurements is much higher than predicted, which strengthens the assumption that the high in-band noise comes from the high jitter and the limited data generator memory. The measurements of the single-loop modulator show higher spurious tones and second-order tones as was predicted by the simulations.

Measurements have been performed at the fractional division numbers of the simulations in Appendix B. For a fractional division by 67.577, no fractional spurs are present in the phase noise spectrum, as was predicted by simulations. The measured in-band noise was again higher than predicted due to the limited data generator memory and the high data generator output jitter. From the discussion in Appendix B, it is clear that $\Delta\Sigma$ fractional-N synthesizer are most sensitive to fractional spurious tones close to integer multiples of the reference frequency. Therefore, measurements have been performed for a fractional division by 67.9923, i.e. 200 kHz from an integer multiple of f_{ref}, which is the worst-case situation in terms of spurious suppression. The measurements are presented in Fig. 6.32. The measured fractional spurs at 200 kHz are below -50 dBc for the MASH modulator. For the single-loop modulator the fractional spurious tones were a few dBs higher and more harmonics appeared in the spectrum, as predicted by the simulations. Also a sub-harmonic shows up at 50 kHz. The overall phase noise levels were equal to those of the other measurements. In Fig. 6.32, the measurements are compared to the spurious suppression mask for the DCS-1800 standard (see Fig. 2.24). The spurs are at least 20 dB below the mask for the MASH $\Delta\Sigma$ modulator and 10 dB for the single-loop $\Delta\Sigma$ modulator. The presented fractional-N synthesizer complies with the DCS-1800 spurious specification even for worst-case frequencies close to an integer multiple of f_{ref}.

The flexibility of the measurement setup enabled the explorations of different timing schemes. A simple but effective rule of the thumb is to perform all noise sensitive signal processing in an as quite as possible environment, i.e. minimize the "digital feedthrough" to sensitive processes. In the case of a $\Delta\Sigma$ fractional-N synthesizer, this comes down to separating the moment when the charge pump pulses are generated from the computational events in the $\Delta\Sigma$ modulator and the prescaler. In this implementation, the PFD and the charge pumps are triggered by the falling edge of the reference clock while the $\Delta\Sigma$ modulator and consequently the prescaler are triggered by the rising edge of the reference clock. To further reduce the "digital feedthrough", the delay between the modulator clock and the PFD clock is swept over a full reference clock period. For large delays between 5 and 10 ns, i.e. the largest possible delay between two edges for a 26 MHz reference clock, the timing between the $\Delta\Sigma$ modulator and the PFD was lost, such that the synthesizer was unable to synthesize the wanted output frequency. For small delays, up to 5 ns, no difference was observed in both the in-band and out-of-band phase noise measurements. One would expect that a small delay, separating the falling edges of the $\Delta\Sigma$ modulator clock and the PFD clock, would further reduce substrate and power supply line noise coupling; This is however not observed due to the noisy (i.e. high jitter) measurement environment. Another timing issue is whether the $\Delta\Sigma$ modulator is clocked by the frequency divider output or the reference clock. Different published implementations of $\Delta\Sigma$ fractional synthesizers use both timing schemes, but the clocking by the frequency divider output is more popular. Both options

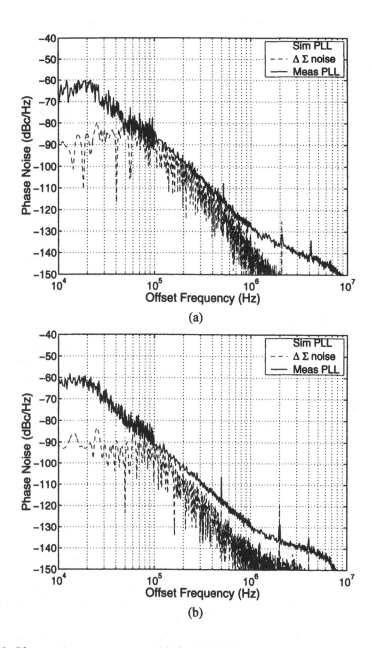

Figure 6.33: *Phase noise measurement with the (a) MASH and (b) the single-loop $\Delta\Sigma$ modulator at 1.76592 GHz compared to the simulated $\Delta\Sigma$ noise at the output of the PLL (dashed) and with the simulated PLL output without $\Delta\Sigma$-control (dash-dotted).*

	DCS-1800	Measured
Phase Noise	−116	< −120
at 600 kHz (dBc/Hz)		< −124 (N=integer)
Phase Noise	−133	< −139
at 3 MHz (dBc/Hz)		< −142.5 (N=integer)
$\Delta\Phi_{rms}$	2°	3° (1.7° for N=integer)
Settling Time (μs)	865/288	215
Reference Spurs (dBc)	−80	< −75
Fractional Spurs (dBc)	−80	< −100
	see Fig. 6.32	
Power Consumption (mW)	–	70
Power Supply (V)	–	2
Frequency Resolution (Hz)	200 kHz	400 Hz

Table 6.2: *Summary of measured specifications compared to the DCS-1800 specifications of Table 2.2.*

were tested, revealing no difference in the measured phase noise spectrum and spurious tones. A possible advantage of clocking the $\Delta\Sigma$ modulator with the frequency divider output is better reference spur suppression; Since in a $\Delta\Sigma$ fractional-N synthesizer, the loop is never locked, the frequency divider output is never exactly equal to f_{ref}. Clocking with the divider output therefore injects less energy at f_{ref} in the substrate, reducing the reference feedthrough. This was however not observed in the measurements.

The general performance of the PLL is the same as in Chapter 5; The measured settling time of the PLL is 226 μs for a 104 MHz frequency step and for a settling accuracy of 100 Hz and thus 215μs for DCS-1800 (see Chapter 5). The power consumption of the PLL is 70 mW from a single 2 V power supply. The IC area is 2×2 mm^2, including bonding pads and bypass capacitors. Table 6.2 shows the measured specifications compared to the DCS-1800 specifications, derived in Chapter 2. The measured phase noise performance complies with the stringent DCS-1800 for integer as well as for fractional division, even at 3 MHz offset, which is the most critical offset frequency for DCS-1800. The measured rms phase error for fractional operation is with 3° worse than the postulated 2°, but still sufficiently below the overall DCS-1800 specification of 5° (Section 2.6.3.4). The measured rms phase noise is higher than expected due to the limited data generator memory and the high data generator jitter. Simulations predict that the in-band phase noise of the $\Delta\Sigma$ fractional-N PLL is not much influenced by $\Delta\Sigma$ noise for moderate non-linearities in the loop, such that the actual rms phase error of the synthesizer is lower than 1.7° The settling performance complies with high data-rate communication demands. The fractional spur level is well below -100 dBc. The reference spur level is lower than -75 dBc, which slightly higher than the DCS-1800 specification. The reference spurs measured for the integer PLL are however below -84 dBc.

The frequency resolution is with 400 Hz much higher than the required 200 kHz , due to the 16-bit internal accuracy of the $\Delta\Sigma$ modulator. This high accuracy was not chosen for the

requency resolution but to ensure enough randomization of the spurious energy in the $\Delta \Sigma$ mod-
lator output. The power consumption is rather high due to fully integrated low-phase-noise
/CO, which is responsible for more than 66% of the total power consumption. The total syn-
hesizer is designed to operate at a power supply of 1.8V. However, at this low power supply
'oltage, the voltage range at the output of the filter adder is not sufficient to synthesize the full
requency range of DCS-1800. Therefore, a power supply voltage of 2V has been chosen.

In short, a low-voltage, 1.8 GHz $\Delta \Sigma$-controlled fractional-N frequency synthesizer is imple-
nented in a 0.25 μm standard CMOS technology. The $\Delta \Sigma$ control of the prescaler moduli is
.cheduled for the digital baseband chip of the fully integrated analog transceiver front-end. The
.pecifications of the IC prototype comply to the DCS-1800 specifications, in spite of the memory
imitations and high jitter of the measurement setup, which degrades the rms phase error and the
eference spurious suppression. The fractional spurious performance of the fractional-N synthe-
iizer complies with the DCS-1800 spurious specification even for synthesized frequencies close
o an integer multiple of f_{ref}.

5.9.3   Comparison with Published $\Delta \Sigma$ Fractional-N Synthesizers

[o demonstrate the quality of the presented $\Delta \Sigma$ fractional-N synthesizer, this work is compared
o previously published $\Delta \Sigma$ fractional-N synthesizers in Table 6.3. The main advantage of frac-
ional synthesis is the decoupling of the reference frequency and the loop bandwidth, enabling
ntegration of the loop filter and a faster settling. The main bottleneck is the spurious perfor-
nance. The key comparison factor is the degree of integration, since a higher integration level,
which is facilitated by fractional synthesis, makes it harder to obtain reasonable specifications,
especially in lossy CMOS. Moreover, external components enable tuning of the loop filter bandwidth (and f_{ref}) and to try different VCO implementations to obtain an optimal performance.

Other important implementation parameters are the $\Delta \Sigma$ implementation, the chosen f_{ref} and
loop bandwidth f_c. Next, the in-band noise is listed, since noise leakage could corrupt the
spectral purity due to the non-linearities of the building blocks. A higher loop bandwidth also
paves the way for lower in-band noise. The spectral purity, including fractional and reference
spurs, is compared. The phase noise at 3 MHz offset is compared, since this is the hardest
specification for DCS-1800. The published phase noise performance is scaled to 1.8 GHz and if
necessary extrapolated to 3 MHz offset, to allow a fair comparison. Last but not least, the settling
performance is shown in μs /MHz to normalize the settling towards the imposed frequency step.
It is however strange that only a minority of the published works mention the settling time,
although it can be greatly enhanced by the fractional synthesis.

The power consumption of the different works is not compared since this is extremely dependent on the degree of integration of the synthesizer. The power consumption of the digital
part of the synthesizer (all building blocks, except the VCO and the loop filter) is always of the
same order of magnitude: between 20 and 30mW. The presented fractional-N synthesizer's digital power consumption is 16.2mW, including the frequency divider, PFD, dual charge pumps
and linearization and timing circuits. An integrated $\Delta \Sigma$ modulator consumes maximally 2mW,
especially when pipelining is employed [Perr97]. Consequently, the power consumption of the
presented work is on the low end of the published fractional-N synthesizer spectrum.

	Implementation	ΔΣ	f_0	f_{ref}	f_c	In-band (dBc/Hz)	Spurs (dBc)	$\mathcal{L}$(3MHz) (dBc/Hz)	Settling (μs /MHz)
[Mil91]	Discrete	3rd MASH	750MHz	200kHz	750Hz	−70	−133	−142	NA
[Rii93]	Discrete	3rd 1b (PC)	405MHz	10MHz	30kHz	−85	−82	−137	NA
[Per97]	0.6μm CMOS ext. /2, VCO	2nd MASH	1.8GHz	20MHz	84kHz	−75	<−60	−130	NA
[Fii98]	CMOS ΔΣ BJT PLL ext. LPF VCO	4th MASH	915MHz	20MHz	100kHz	<−90	<−90	−133	50/8.4
[Wil00]	0.5μm CMOS ext. LPF VCO	3rd 3b	2.5GHz	48MHz	700kHz	−100	<−77	−123	5/110
[Rhee00a]	0.35μm BiCMOS ext. LPF VCO	3rd 3b	2.5GHz	8MHz	35kHz	−82	<−85	−136	150/?
[Aho00]	0.35μm CMOS ext. LPF VCO	3rd MASH	2GHz	8MHz	35kHz	−79	<−72	−125	NA
[Rhee00b]	0.5μm CMOS ext. LPF VCO	2nd/3rd MASH	1.1GHz	7.994MHz	40kHz	−92	<−90	−131	NA
[Lee01]	0.5μm BiCMOS ext. LPF VCO	4th 1b	1.643GHz	9.84MHz	12kHz	−87	<−68	−143	500/30
[Lo02]	0.5μm CMOS	3rd MASH	900MHz	25.6MHz	80kHz	NA	<−67	−128	100/?
This work	0.25μm CMOS	3rd MASH 3rd 4b (DG)	1.8GHz	26MHz	35kHz	−73 −82	−75 −84	−139 −142.5	226/104

Table 6.3: *Comparison of the presented ΔΣ fractional-N synthesizer with previously published ΔΣ fractional-N synthesizers.*

The first two publications in Table 6.3 ([Mill91], [Ril93]) are focused on proving the feasibility of $\Delta\Sigma$ fractional-N synthesis. [Perr97] is in fact pioneering for CMOS $\Delta\Sigma$ fractional-N synthesis and transmission and presents good measurement results. The VCO and the high-speed divide-by-2 flipflop are external. The 5 following publications all implement only the digital synthesizer part, without VCO and loop filter. The most eye-catching publications are [Will00] and [Rhee00a]. The first one presents a different phase detection scheme and a high bandwidth to enable super fast frequency hopping. The second one has chosen a much smaller bandwidth, resulting in an excellent spectral purity, but the VCO and the loop filter are tuned to yield the best noise performance. Only the last implementation [Lo02] presents a fully integrated solution with a digitally controlled VCO and integrated loop filter; The spurious suppression and the out-of-band phase noise is moderate and the in-band noise is not even mentioned. The settling time is as fast as $100\mu s$, but the corresponding frequency step is not mentioned in the text.

When the presented 1.8 GHz $\Delta\Sigma$ fractional-N synthesizer is compared to the previously published ones, its main contribution is an excellent spectral purity combined with a fully integrated solution. Only the $\Delta\Sigma$ modulator is not integrated on the die. In Table 6.3, the measured spectral purity is given for fractional measurements (on top) and integer measurements. When this work is compared with other implementations, one thing is striking: the rather small loop bandwidth of 35kHz. $\Delta\Sigma$ fractional-N synthesis sets the trend toward higher loop bandwidths (in the order of 100 kHz) for faster loop dynamics, higher integratability and lower in-band noise. Here, a loop bandwidth of 35 kHz is chosen, since the toughest specification of DCS-1800 is not the in-band noise specification but the out-of-band noise specifications at 600 kHz and 3 MHz offset. A higher loop bandwidth allows noise from the loop filter pole τ_4 and the active loop filter circuits to appear at these offset frequencies, degrading the carefully optimized phase noise of the integrated VCO. What's more, the linear analysis of Section 6.6 showed that the maximum loop bandwidth is only 62 kHz in order to comply with the DCS-1800 specifications. In Table 6.3, the only implementation with roughly the same phase noise performance has a loop bandwidth of only 12 kHz, resulting in slow settling and an off-chip loop filter.

The measured performance of the presented $\Delta\Sigma$ fractional-N synthesizer is competitive with respect to the other published fractional-N synthesizers, but –apart from one work [Lo02]– it presents the only fully integrated solution. The out-of-band noise of the presented synthesizer is almost the lowest ever published and exceeds that of the other fully integrated fractional-N synthesizer [Lo02]. The spurious suppression and the in-band phase noise are comparable, especially when the memory limitations and high jitter of the measurement setup are taken into account. The settling performance is the best published, apart from the one in [Will00], which was designed for fast settling, but suffers from high out-of-band noise. The presented $\Delta\Sigma$ fractional-N synthesizer is to the my knowledge the only fully integrated fractional-N synthesizer with an excellent spectral purity, complying with the DCS-1800 standard. The $\Delta\Sigma$ modulator is not integrated on the die[3]. Integration of the $\Delta\Sigma$ modulator would alleviate the need for the performance limiting, high jitter measurement setup, resulting in measurements close to the integer ones, with better spurious suppression and lower in-band noise, exceeding the DCS-1800 specifications.

---

[3]As stated earlier, integration of the $\Delta\Sigma$ modulator was planned, but not finished due to technology access problems.

6.10   Conclusion

This chapter presents a thorough analysis of $\Delta\Sigma$ fractional-N synthesizers, linear as well as on-linear, which has led to the implementation of a monolithic 1.8 GHz $\Delta\Sigma$-controlled fractional-N PLL frequency synthesizer in a standard 0.25 μm CMOS technology, starting from the PLL of Chapter 5. As stated earlier, $\Delta\Sigma$ fractional-N synthesizers are very well fit for integration in CMOS; The capability of fractional division decouples the frequency resolution and the loop bandwidth, leading to higher integratability and faster dynamics. Moreover, the $\Delta\Sigma$ fractional-N synthesizer employs the same, simple architecture as a PLL extended with digital logic, to trade frequency resolution against digital complexity, i.e. the biotope of CMOS.

Since $\Delta\Sigma$ modulators generate quantization noise and pattern noise for non-busy inputs, the influence of the $\Delta\Sigma$ modulator on the spectral purity of the synthesizer needs to be investigated. The 3rd-order MASH (cascade 1-1-1) and multi-bit, single-loop $\Delta\Sigma$ topologies are chosen as guinea pigs. First, the $\Delta\Sigma$ influence is analyzed theoretically using linear system theory; Under the constraint of the DCS-1800 noise specs, the maximum PLL bandwidth is calculated versus the reference frequency, i.e. the raison d'être of fractional-N synthesis. For a reference frequency of 26 MHz, the in-band noise sets the maximum bandwidth to 810 kHz for the MASH and 614 kHz for the single-loop $\Delta\Sigma$ modulator. However, the more stringent out-of-band noise spec restricts the bandwidth to only 87 kHz for the MASH and 62 kHz for the single-loop modulator. According to the linear analysis, the introduction of a $\Delta\Sigma$ modulator in the PLL only limits the possible bandwidth for low out-of-band phase noise but poses no threat to the in-band noise (the rms phase error) nor to the spurious suppression.

Practice however proves the theory wrong; To quantify the real fractional behavior, a fast, non-linear analysis method is developed. The method takes into account the non-linear $\Delta\theta \rightarrow I_{qp}$ conversion (the PFD and charge-pumps) with two parameters: a dead zone in the PFD and gain mismatch in the charge pumps. In contrast to the theoretical analysis, the non-linear analysis showed severe in-band noise leakage and emerging spurious tones at the fractional frequencies. Due to the high simulation speed, the analysis can be swept over different degrees of non-linearity and for different input values, revealing the following observations:

- The $\Delta\Sigma$ phase noise at 600 kHz is lower for a MASH modulator in the ideal case, but only a small non-linearity is enough to favor the single-loop modulator.

- The phase noise at 3 MHz is more critical for a single-loop modulator, due to the pole shift in the noise transfer function. The noise at 3 MHz is virtually not influenced by non-linearities.

- The in-band noise leakage for a single-loop $\Delta\Sigma$ modulator is always lower than for a MASH modulator. For the MASH modulator, noise leakage is highly critical in the presence of a dead zone.

- The influence of a very small dead zone is much more severe than the influence of gain mismatch. A dead zone in the $\Delta\theta \rightarrow I_{qp}$ conversion must be avoided at all times.

- The on-time of the PFD is higher for the MASH than for the single-loop modulator, which increases the sensitivity of the synthesizer to digital noise coupling through the substrate and the power supply.

- The value of the fractional division input number has no direct influence on the noise leakage due to non-linearities.

- For fractional division by numbers in between integer division moduli, no spurious tones are present in the output spectrum of the synthesizer.

- The contrary is true for fractional division close to integer division moduli. Spurious tones show up at the synthesizer output at the fractional frequencies, although dithering is applied. The spurious tones are higher when the synthesized frequency approaches an integer multiple of the reference frequency.

To realize the fractional-N synthesizer, the 4th-order, type-II PLL of Chapter 5 is extended with a $\Delta\Sigma$ modulator, with a focus on the design of the $\Delta\theta \rightarrow I_{qp}$ conversion blocks. A zero-dead-zone PFD is implemented by a reset delay; The delay must be high enough to suppress the dead zone, but as small as possible to reduce the sensitivity to noise coupling and high frequency decision errors. The design of the charge pumps is optimized to minimize the spurious tones by a custom timing scheme and dummy branches. The charge pump current derivation is carefully implemented, leading to a maximum gain mismatch of ±0.5%.

Since in this implementation, the $\Delta\Sigma$ modulator is scheduled for integration in the digital baseband IC of the transceiver system, the measurements are performed by an externally applied $\Delta\Sigma$ bitstream. The accuracy of the bitstream (especially in terms of jitter) is limited by the measurement setup, leading to an increased phase noise at the smallest offset frequencies. Apart from this offset region, the measurements and simulations show good correspondence. The measured phase noise is lower than -120 dBc/Hz at 600 kHz and -139 dBc/Hz at 3 MHz. The specifications of the IC prototype are close to the DCS-1800 specifications, in spite of the limitations of the measurement setup, which degrades the rms phase error and the reference spurious suppression. The fractional spurious suppression satisfies the DCS-1800 spurious mask even for synthesized frequencies closest to an integer multiple of f_{ref}. The measured phase noise of the single-loop modulator is lower than that of the MASH modulator, with higher spurious tones as predicted by the simulations. Although the MASH is easy to integrate and stable, the single-loop modulator provides more flexibility and lower noise and less sensitivity to noise leakage and noise coupling.

To demonstrate the quality of the presented $\Delta\Sigma$ fractional-N synthesizer, the work is compared to previously published $\Delta\Sigma$ fractional-N synthesizers; The presented fractional-N synthesizer has among the lowest out-of-band phase noise values reported in integrated CMOS synthesizer designs combined with fast loop dynamics for a very high degree of integration; The presented synthesizer is the only one that presents a fully integrated PLL solution in CMOS with high specifications. As derived from the simulations, it is believed that integration of the $\Delta\Sigma$ modulator on the same die would further improve the measured performance of the synthesizer[4]; The digital noise coupling of the 26 MHz $\Delta\Sigma$ modulator on a high-ohmic substrate with guard ring isolation is unlikely to be worse than the noise introduced by the employed measurement setup. Therefore, by integrating the $\Delta\Sigma$ modulator on the same die, which would

---

[4]The integration of the digital $\Delta\Sigma$ modulator is not a highly complex task and was intended for the next version of the DCS-1800 frequency synthesizer. The final integration step could not be accomplished since access to the 0.25μm technology, which was licensed by Toshiba, was denied.

only marginally increase the power consumption, the performance of the presented fractional-N synthesizer would exceed the DCS-1800 requirements, while presenting a single-chip solution, without external components or technological tours de force.

Chapter 7

Conclusions

7.1 A 2V CMOS Cellular Transceiver Front-End

The research presented in this book which led to the monolithic integration of a DCS-1800 compliant, $\Delta\Sigma$-controlled fractional-N synthesizer in CMOS, is part of a larger picture: the monolithic integration in CMOS of a low-voltage transceiver front-end that meets the stringent specifications of the class I/II DCS-1800 standard. The research was ultimately successful with as highlight the presentation at the 2000 International Solid State Circuit Conference (ISSCC'00) of the first ever published CMOS monolithic transceiver front-end with cellular specifications [Stey00b, Stey00a]. The transceiver combined for the first time a fully integrated low-IF receive path (LNA, I/Q mixers and VGAs), a direct-upconversion transmit path (I/Q mixers and output stage) with a complete *on-chip* fractional-N PLL, with an integrated quadrature VCO, 64/79 prescaler, loop filter and phase detector, on a single 0.25μm CMOS die. Moreover, operating from a single 2V power supply voltage, it was the *lowest voltage CMOS RF front-end* published with cellular specifications. In what follows the transceiver front-end building blocks are briefly discussed. A more elaborate discussion is given in [Jans02, RX] and [Borr02, TX].

The low-IF RX consists of an LNA connected to two (I&Q) down-conversion mixers followed by a variable gain amplifier-filter (Fig. 7.1, bottom-left). The front-end converts the RF signal into a differential I and Q signal, centered at an 100 kHz IF. The cascode LNA is input-matched using inductive source degeneration through an on-chip spiral inductor. The measured S_{11} is better than -11.5 dB between 1.725 GHz and 1.975 GHz, satisfying the antenna filter requirement ($S_{11} < -10$dB). The lower input impedance improves the LNA gain by about 1.5 dB compared to an exact 50Ω match. The on-chip inductor at the LNA output resonates with the load capacitance of the I and Q mixers, centering the gain at 1.84 GHz. Each down-conversion mixer makes use of a cascoded, current folding switching mixer. The folding mechanism allows low voltage operation while enabling the insertion of a cascode transistor to reduce LO leakage and increase mixer linearity. The large $V_{GS} - V_T$ of the input transconductor sets $IP3$ to +17 dBVref$_{[224mVrms]}$. The noise contribution of the top current source is reduced by supplying part of the current of the input transconductor through a high $V_{GS} - V_T$ PMOS bleeder. A CMFB-circuit sets the common mode level on the interface between mixer and VGA to 1 V. The VGA consists of a two-stage fully differential OTA with a bank of highly matched, RC transimpedance elements (8R1C,8R1C,4R2C,2R4C,1R8C). Its gain can progressively be decreased by 6 dB by switching them in parallel. The RC transimpedance also performs first order filtering of the blocking signals. Due to the large loop gain, the output of the VGA stays very linear until the 1 dB compression point of +15 dBVref$_{[224mVrms]}$ is reached. A/D conversion, channel selection, mirror suppression and additional AGC is assigned to a separate digital CMOS base-band chip. The RX gain (max. VGA) increases from 54.3 dB at 1.8 GHz to 54.5 dB at 1.83 GHz, and decreases down to 53.3 dB at 1.88 GHz. The worst-case image rejection ratio (IMRR) of the RX path is 32.2 dB at 200 kHz offset from the LO. The DC offset due to LO self-mixing and mismatch at the highest RX (VGA) gain level is 1.1 Volt (6.9 dBVref$_{[224mVrms]}$). Since a low-IF RX does not employ the information at DC, the DC offset can be accounted for by putting an extra 1/2 bit in the A/D. The total noise figure (NF) of the complete Low-IF RX path, integrated in the 200 kHz band around the 100 kHz IF (I+jQ) is 8.2 dB, of which 5.6 dB is caused by white noise and 4.7 dB by 1/f noise. The $IIP3$ of the RX is better than -6.2 dBm for all VGA gain settings. The net LO leakage into the antenna is less than -63 dBm in the DCS band. The RX signal path consumes 113 mW at 2 Volt, of which the LNA is taking up only 10 mW.

The upconversion is performed by a quadrature linear transconductance mixer (Fig. 7.1, top-right). The mixer transistors are dimensioned to achieve better than 33 dB mirror suppression. Further increasing the mixer transistor width improves mirror suppression and signal amplitude. However, since a larger load capacitance requires smaller resistors in the polyphase circuit, care must be taken not to degrade the VCO phase noise. Therefore, the mixer transistor dimensions have been optimized to achieve an optimal global transmitter performance, including the VCOs phase noise and the global power consumption. This illustrates that the system is designed as one large transistor level circuit where global system performance prevails on building block speci-fications. The mixers are followed by a linear, low input impedance output buffer. The output driver consists of a two-stage folded NMOS-mirror structure with active LF signal suppression and DC biasing. The buffer consumes 12 mA and the output driver 29 mA. The linearity only depends on the matching properties of the devices, the gate overdrive and the AC to DC current ratio. Because of the relatively large transistor width a low DC/AC current ratio can be allowed. AC coupling in the buffer allows to set the baseband DC voltage independently of the V_{GS} of the current mirror. The small node impedance keeps the signal excursions small. The DC-feedback in each stage sets the bias voltage of each stage to the same value, optimizing the linearity of the current mirrors. The folding of the RF signal has the advantage that the power supply is not affected by the large RF currents. In this way coupling of the RF signal to the VCO which can deteriorate phase noise performance, is avoided. Both the up- and down-conversion mixers are steered by a quadrature LO signal. The quadrature oscillator is optimized as one building block, trading-off output signal and VCO phase noise against global power consumption and avoiding excessive power consumption due to the use of intermediate buffers. The output buffers to the up- and down-conversion mixers share the same 1 mA current source while the cascode transistors select either the up- or down- converter. The switch transconductance is about 10 mS. The GMSK output spectrum at 1.8 GHz, measured at 30 kHz RBW, is compared to the ETSI DCS-1800 mask [ETSI00]. The measured noise at large distance is smaller than -78dBc (RBW 100kHz, Cfr. ETSI), also complying to the ETSI specs. The measured noise floor is in

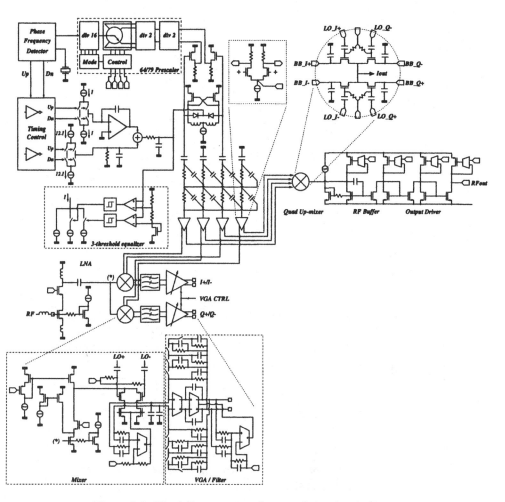

Figure 7.1: *The full transceiver front-end circuit schematic.*

fact limited by the noise of the baseband quadrature generator. The mirror suppression is better than -33 dBc and the LO feedthrough and all spurious distortion components are below -38 dBc. The measured absolute output power is -13 dBm in a non-ideal 50Ω load, i.e. including board parasitics, the SMA connectors, cable and the input-impedance of the spectrum analyzer. The LO signal for both RX and TX path is generated by a monolithic ΔΣ-controlled fractional-N frequency synthesizer, presented in this book (Fig. 7.1, top-left). A chip microphotograph of the 15.4 mm2 IC is shown in Fig. 7.2 (a). The total power consumption of the 2 V, 0.25 μm CMOS transceiver chip is 191 mW (113+70+8) in RX mode and 160 mW (82+70+8) in TX mode. The TRX measurement results are summarized in Fig. 7.2 (b).

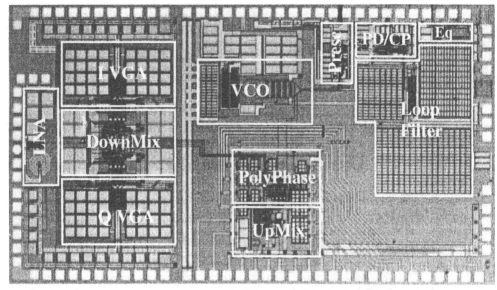

(a)

	Specification		TRX	DCS-1800
	NF	dB	6.2	8.9
	IMRR	dB	32.2	32
Low-IF	Gain	dB	54.5	– –
RX	S_{11}	dB	<-11.5	-10
	IIP3	dBm	>-6.2	-23.5
	LO Leakage	dBm	<-63	-47
	Power	mW	113	ALAP
	Settling Time	μs	226	288
	PN at 600 kHz	dBc/Hz	-120 /-125	-116
$\Delta\Sigma$	PN at 3 MHz	dBc/Hz	-139 /-144	-133
Fractional-N	$\Delta\Phi_{RMS}$	o	3 /1.7	2
PLL	Reference Spurs	dBc	-75 /-84	-80
	Fractional Spurs	dBc	Compliant[a]	ETSI
	Power	mW	70	ALAP
	Output Power	dBm	-13	0
Direct	Mirror Suppression	dBc	-33	-30
Upconversion	LO Feedthrough	dBc	-38	-30
TX	GMSK Mask		Compliant	ETSI
	Power	mW	82	ALAP

[a] see Fig. 6.32 and Section 2.6.3.3

(b)

Figure 7.2: *(a) The microphotograph of the full transceiver front-end. (b) A summary of the measured building block specifications compared to the DCS-1800 standard.*

7.2   Main Contributions and Achievements

The presented research fits in the quest for a fully integrated, CMOS-only transceiver front-end, which meets the stringent cellular specifications, in this case DCS-1800. Full integration has been chosen to reduce the number of components in cellular handsets and consequently the cost. To further lower the cost, the cheapest technology has been selected: CMOS. Despite the inherent lower performance of CMOS, the goal of this book is to prove the feasibility of high specifications in standard CMOS without external components, tuning or trimming.

A major challenge in the design of CMOS single-chip transceiver front-ends is the frequency synthesizer. Its spectral purity is critical for the quality and reliability of the information transfer. Its dynamic behavior determines the maximal information throughput, which is of utmost importance for 2.5G and 3G systems. This book presents the design trajectory of a monolithic frequency synthesizer in standard CMOS technology, focused on attaining the highest spectral purity and fastest switching with moderate power consumption. The design trajectory goes from the specification derivation down to the systematic development of different CMOS circuits, with ultimately the successful implementation of a monolithic frequency synthesizer.

Architectural exploration has led to the choice of a $\Delta\Sigma$ fractional-N PLL, since it trades off accuracy (i.e. high specifications) for digital complexity, i.e. the natural biotope of CMOS. By carefully interpreting the DCS-1800 requirements [ETSI00], measurable synthesizer specifications have been extracted. The relative importance of the different building blocks on the overall synthesizer performance has been investigated and used to set their requirements. The translation of the derived building block specifications onto actual circuits down to transistor level forms the more theoretical core of the presented book. Insights on $\Delta\Sigma$ fractional-N PLL and PLL building block design are gathered and design guidelines are given.

- In the quest for the fastest divide-by-2 flipflop, a systematic sizing strategy is employed based on optimization of the circuit for divide-by-2 operation and careful design of each circuit node for speed and power. Moreover, design rules are developed to design the prescaler circuits for low residual phase noise and high input sensitivity.

- A digression is made onto the exploration of the speed limitations of conventional CMOS. By proper circuit design, a CMOS prescaler is shown to operate at above-12 GHz, even in first-generation 0.25μm technologies.

- For VCO design a design-oriented non-linear phase noise theory is presented that not only takes into account the noise of the passive and active devices, but also the influence of the non-linear active element. The theory was compared to existing phase noise theories. A simulator-optimizer program is developed to realize optimal-Q inductor geometries in CMOS to reduce white phase noise generation.

- Moreover, 1/f noise upconversion mechanisms have been identified and successfully countered in the VCO circuit design, by choosing the operating point that maximizes waveform symmetry. A bias filtering technique to remove all flicker noise that also significantly reduces white phase noise has been elaborated and experimentally verified.

- The integrated PLL is theoretically analyzed for noise and capacitance; A systematic design strategy has been developed that marks out the design space, enabling fully integrated loop filters.

- The influence of $\Delta\Sigma$ modulator noise in fractional-N PLLs is theoretically analyzed and the loop bandwidth limitations are elaborated for phase noise and rms phase error.

- A fast non-linear analysis method has been developed to more accurately predict the effect of $\Delta\Sigma$ modulators in fractional-N PLLs; Non-linearities, i.e. a dead zone and gain mismatch, in the PFD and charge pumps are identified as the main sources of in- and out-of-band $\Delta\Sigma$ noise leakage and spurious tones. The simulation results were in close correspondence with the measurements.

The research performed in this book has led to several CMOS implementations of PLL building blocks and a fully integrated frequency synthesizer.

- A single-ended 1.5 GHz 8/9 dual-modulus prescaler in 0.7μm CMOS with optimization of the residual phase noise and high input sensitivity [DeMu98].

- A single-ended 1.8 GHz 8/9 dual-modulus prescaler in 0.8μm BiCMOS with a custom ratio-ed logic D-flipflop to maximize the speed-power ratio with a high input sensitivity.

- A 12 GHz /128 prescaler in a first-generation 0.25μm CMOS which features a divide-by-2 flipflop, operating up to 15 GHz [DeMu00c].

- A 2 GHz low-phase-noise integrated LC-VCO set in 0.65μm BiCMOS with circuit tricks to minimize the flicker noise upconversion. Phase noise is as low as -125dBc/Hz at 600 kHz and the $1/f^3$ phase noise corner is reduced to 15 kHz [DeMu00b, DeMu99a].

- A low-impedance bias current source filter technique for VCOs is implemented and experimentally verified; Due to the filtering, the VCO has no flicker noise upconversion and phase noise as low as -132.5 dBc/Hz at 600 kHz. This measured performance exceeds that of all published fully integrated CMOS VCOs, with and without filtering.

- A 1.8 GHz fully integrated VCO in a first-generation 0.25μm CMOS process with a tuning range of 28% and phase noise as low as -127.5dBc/Hz at 600 kHz, which was the best ever reported fully integrated CMOS VCO at that time in open literature [DeMu00a].

- A 2V, 1.8 GHz fully integrated $\Delta\Sigma$-controlled fractional-N PLL in 0.25μm CMOS. The IC includes a fully integrated 35 kHz dual-path loop filter, a fully integrated LC-tank VCO, a high-speed 64/79 prescaler, a zero-dead-zone phase-frequency detector, a 3-step equalizer and dual charge pumps. Its phase noise is as low as -125.7 dBc/Hz (integer) [DeMu00d, Stey00b] and -120 dBc/Hz (fractional) at 600 kHz and the frequency resolution is 400 Hz, with only 1.4 nF on-chip. The fractional spurious suppression complies with the DCS-1800 standard even for fractional frequencies close to integer multiples of the reference frequency [DeMu02].

- The VCO of the presented fractional-N synthesizer is capable of driving a polyphase filter in a power efficient manner [Borr00]. These realizations made it possible to integrate a complete 2V cellular transceiver front-end in CMOS [Stey00a], without trimming, tuning and external components. The PLL operates from a power supply voltage of 2V to prove the feasibility of low-voltage RF CMOS for compatibility with the digital part in view of single-chip transceiver integration.

7.3   Epilogue

During the six years of my Ph.D. research, I could witness the gradual turnabout in the believe of the micro-electronics world in monolithic RF-CMOS. At first, only some "pigheaded" people believed in the future of CMOS for any kind of RF applications, certainly for monolithic integration, although pioneering work was already performed, with the MICAS-group of the K.U.Leuven as one of the most important trendsetters. Gradually, this point of view shifted towards acceptance of RF-CMOS by researchers as well as industry, albeit mostly for low-end applications and for low degrees of integration (i.e. with external components). This was reflected in the booming number of RF-CMOS-publications (and the ever increasing, tedious review work for RF-CMOS research assistants). Today, fully integrated RF-CMOS is starting to gain the respect it deserves, even for cellular specifications with the emergence of commercial, monolithic RF-CMOS transceivers (front-ends) for low- [Dar01, Eynd01] and high-end applications (3 chip CMOS GSM chip-set of SiLabs [Aero01, Fren01]).

I like to believe that the work of my RF-predecessors and my RF-colleagues has contributed to accomplish this turnabout and that we proved the disbelievers wrong, what does not mean that the work to make monolithic RF-CMOS an industrial standard is finished. Anyway, since all is cost-driven, I strongly believe that the degree of integration will keep on increasing and RF-CMOS will keep on spreading, especially with the ongoing effort put in tuning standard CMOS technologies for higher RF performance.

Some dissonant voices might say that ,due to the continuous CMOS scaling, digital will push the need for analog and RF design aside (i.e. the software radio paradigm), but again I disagree. Today, the trend is to put more and more complexity and functionality into a single chip, the so-called SoCs, and to go to higher processing speeds. Although at first sight SoC design might seem to even further push analog away, the opposite is true: With the high-complexity and higher speeds, more and more analog effects start to emerge such as cross-talk, leakage currents, networks-on-chip, DC and speed control,... This means that digital is becoming more and more analog, which is even confirmed by digital gurus such as Jan Rabaey [Rab02].

With the --more and more analog-- future in mind, I hope that the presented research has proven its validity and usefulness and will become yet another --albeit small-- step towards monolithic (RF-)CMOS systems, from which all of mankind could benefit one day.

Appendix A

$\Delta\Sigma$ Modulators with DC-inputs

In $\Delta\Sigma$ fractional-N frequency synthesizers, the $\Delta\Sigma$ modulator must provide a bitstream with a fractional mean value, equal to the value applied at its input. In other words, the input is a constant value, further called a DC value, such that the white noise assumption is not valid. Successive samples of the quantization error $e[i]$ are highly correlated and its power spectral density is no longer flat. As a result, linear analysis of the single-bit, first order $\Delta\Sigma$ modulator is a rough approximation and cannot predict how the power spectral density of the quantization noise will look like, while this is of major importance in synthesizer design.

When looking at the first order $\Delta\Sigma$ modulator, it is clear that in order to approximate the input using only 0 and 1, the output will contain repetition, dependent on the input signal. This manifests itself as large spurious tones in the power spectral density of the $\Delta\Sigma$ modulator output. The quantization noise is said to be colored and is called *pattern noise*. Fig. A.1 shows output spectrum of a single-bit, first order $\Delta\Sigma$ modulator for different DC-inputs. Large spurious tones can be distinguished. In Fig. A.1 (b) the output of the $\Delta\Sigma$ modulator is plotted versus the theoretical output spectrum predicted by Eq. (6.9), given by the solid line. The actual output spectrum contains tones which can be larger than the theoretical quantization noise. To get an idea of what this pattern noise looks like, a top view of Fig. A.1 (a) is plotted in Fig. A.2 (a). The lighter the color at a certain point in the graph, the higher the value of the output spectrum. The white dots denote spurious tones in the $\Delta\Sigma$ modulator output spectrum. Several patterns can be distinguished. First, a large white cross covering the figure is present, which consists of tones, whose frequency is linearly proportional to the input. Secondly, two smaller crosses, symmetrically distributed over the figure, are composed of tones, whose frequency is related in a second order fashion to the input. Less visible but also present are third order frequency related tones. Furthermore, lines parallel to the lines of the large cross are present. At rational inputs, the quantization noise is concentrated in a few discrete tones, giving rise to black vertical lines in the output plot of Fig. A.2(a). Also horizontal white lines can be distinguished indicating tones that are always present, independent of the input.

The precise analytical description of the pattern noise is beyond the scope of this book. It can be found in open literature [Gray89] [Cand81][Eynd93]. However, in order to gain insight in the structure of pattern noise, a summary of the results is given.

Eq. (A.1) shows a mathematical expression for the output of a first order, single bit $\Delta\Sigma$

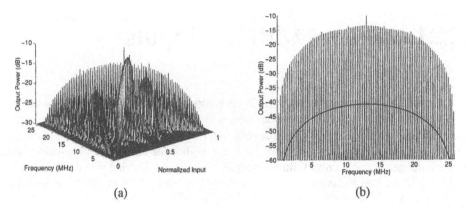

(a)   (b)

Figure A.1: *(a) The output of a first order single bit* ΔΣ *modulator for different DC-inputs.*
Large spurious tones are visible in the frequency spectrum and are referred to as pattern noise.
(b) The output of a first order, single-bit ΔΣ *modulator compared to the ideal* ΔΣ *modulator*
output (the solid line).

modulator with n the input to the ΔΣ modulator, N' the quantization level closest to n (i.e. 0 or
1) and $n' = abs(N' - n)$.

$$y(t) = N' + \sum_l \sum_k \frac{\sin(\pi l n')}{\pi l} \exp(j2\pi \frac{l n' + k}{T} t) \tag{A.1}$$

The frequency of the spurious tones that show up in the output spectrum of the ΔΣ modulator are proportional to the DC-input, i.e. $f_{tones} = (l n' + k) \cdot f_{ref}$. Two different kinds of input
dependency can be distinguished. First, there are tones, whose frequency is l-th order proportional to the input. In Fig. A.2 (a), these tones form the white crosses, the large cross being of
the first order, the two smaller crosses of the 2-nd order, ... Second, tones exist which are offset
from the proportional tones by an integer offset k. These tones form the white lines parallel to
lines of the crosses.

In Fig. A.2 (b), the noise power in the PLL noise band is plotted versus the DC-input to the
ΔΣ modulator for an OSR of 64. The dotted line indicates the theoretical noise power calculated
from Eq. (2.20). The in-band noise power is also dependent of the DC-input and consists of large
peaks which can be a lot higher than theoretically predicted, especially near 0 and 1 and around
rational multiples of the input. In what follows, a more detailed description is given of this noise
pattern.

- When $n = 0$, the output of the ΔΣ modulator consists of all zeros. Therefore the quantization noise is zero.

- When $n = 1/(2p + 1)$, a small DC value with p a large positive integer, the output of the
  ΔΣ modulator will consist mostly of zeros, but some samples will be one. In fact, every

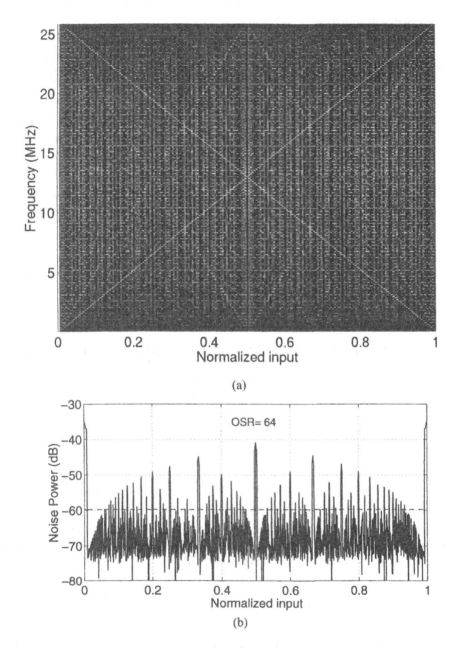

(a)

(b)

Figure A.2: *The pattern noise of a first order $\Delta\Sigma$ modulator for different DC-inputs. (a) A top view of the input-frequency-output power plot. Lighter color indicates a higher output value. The white dots are spurious tones arising from the non-ideal $\Delta\Sigma$ modulator operation. (b) The pattern noise integrated over the noise bandwidth of the PLL. The dashed line indicates the theoretical noise power, calculated from Eq. (2.20)*

$2p + 1$ samples, a one is generated. This means that the quantization noise is colored, i.e. concentrated in discrete tones at frequencies:

$$f_{tones} = kn f_{ref} \quad and \quad k = 1, 2, 3, \ldots \tag{A.2}$$

with a power of:

$$E^2(f) = \frac{2n^2}{\cos^2(\pi f_{tones}/f_{ref})} \approx 2n^2 \tag{A.3}$$

Since it is possible that tones appear inside the noise bandwidth of the PLL, the total rms phase error can be degraded. The number of tones within the bandwidth is :

$$\#_{tones} = \frac{f_{ref}/2OSR}{n \cdot f_{ref}} = \frac{1}{2nOSR} \tag{A.4}$$

As long as there is at least one in-band discrete tone, i.e. $n < 1/(2OSR)$, the noise power within the noise bandwidth of the PLL can be calculated to be:

$$n_{\Delta\Sigma} = E(f) \cdot \#_{tones} = \frac{n}{OSR} \tag{A.5}$$

The equation states that with increasing DC-input n, the noise power increases up to a maximum of $n_{\Delta\Sigma max} = 1/(2OSR^2)$. Simultaneously, the frequency of the tones increases following Eq. (A.2), i.e. k-th order proportional to n and its multiples. If n is further increased no more tones fall within the noise bandwidth and the quantization noise decreases abruptly, as can be seen in Fig. A.2 (b).

- When $n = 1 - 1/p$, a DC value close to one with p a positive integer, the output of the ΔΣ modulator will consist mostly of ones, with every p-th zero. Again, the quantization noise power spectrum is discrete with spectral tones in the noise band. All formulae are analogous to the case where n is close to zero, with n replaced by $n - 1$. The noise power in the noise band now increases when going away from one and abruptly decreases when no spectral tones are present in the noise band. The frequency of the spectral tones is $f_{tones} = k(1 - n) f_{ref}$ and $k = 1, 2, 3, \ldots$ The maximum integrated noise power is given by $n_{\Delta\Sigma max} = 1/(2OSR^2)$.

- When $n = 1$, the output of the ΔΣ modulator is all ones. Therefore the quantization noise is zero.

- In Fig. A.2 (b), the integrated noise power for different DC-inputs is plotted. Large noise peaks are present around 0 and 1, as predicted. In a similar way, noise peaks appear around rational values of n. When n is rational, satisfying $n = q/p$ with p and q integer, positive and incommensurate, two noise peaks are present around n with a maximum value of $n_{\Delta\Sigma max} = 1/2(p \cdot OSR)^2$. The noise power peaks around rational inputs are smaller than the ones at 0 and 1.The noise power at the rational inputs themselves is low and concentrated in a few spurious tones. This is also visible in Fig. A.2 (a), where black vertical lines at rational inputs represent the low in-band noise power values. The much lighter areas around rational inputs indicate that much more noise is present in the noise band.

Appendix B

Additional Results of the Non-Linear Analysis for $\Delta\Sigma$ Fractional-N Synthesizers

For completeness and clarity, this appendix presents additional analysis results of the non-linear analysis method presented in Section 6.7. The analysis method presents a fast way to examine the influence of non-linearities and mismatch in the PLL building blocks on the spectral purity of the $\Delta\Sigma$ fractional-N synthesizer. The main non-linearities that are focused on, are the dead zone in the PFD and the gain mismatch between the up- and the down-currents of the charge pump. In Section 6.7, analysis results are presented for a fractional division by 67.92, i.e. an output frequency of 1.76592 GHz for a 26 MHz reference frequency for MASH modulators as well as for single-loop, multi-bit $\Delta\Sigma$ modulators. The effects of non-linear behavior on the in-band noise and out-of-band phase noise are summarized at the end of Section 6.7.2; Noise leakage, corrupting in- and possibly out-of-band noise, was observed. Another consequence of non-linearities is the emerging of spurious tones, where the $\Delta\Sigma$ modulator output spectrum contains none. For a division by 67.92, spurious tones appear at 2.08 MHz, i.e. $0.08 \times f_{ref}$ and harmonics in the case of a single-loop modulator. The power of the spurious tones is however way below the -80 dBc spurious tone specification imposed by the DCS-1800 telecommunication standard.

The appendix presents additional analysis results to show the influence of the fractional division number on the spectral purity of a fractional-N synthesizer. Two cases are chosen; The first case is a fractional division by a number closest to an integer multiple of the reference frequency in a DCS-1800 system, i.e. by 67.9923, revealing an output frequency 200 kHz from $68 \times f_{ref}$. The second case presents a fractional division by an arbitrarily chosen number in between two integer moduli, e.g. 67.577. Fig. B.1 and Fig. B.3 show the results of an analysis, performed with the following non-linearities: a 0.1% dead zone and a gain mismatch of $\pm 1\%$. The dead zone is again defined in percent of the useful phase range of the phase detector, which is -2π to 2π in the case of a PFD. The size of the introduced dead-zone corresponds to 0.72° around 0° phase error. The gain mismatch is defined as the mismatch between the up and the down current of the charge pumps in $\pm$ percent.

In Fig. B.1, the Matlab-simulated output phase noise of the synthesizer for a division by 67.9923 is presented. Noise leakage can clearly be observed and is higher for the MASH than for the single-loop modulator, as expected from Section 6.7. However, more alarming in this case

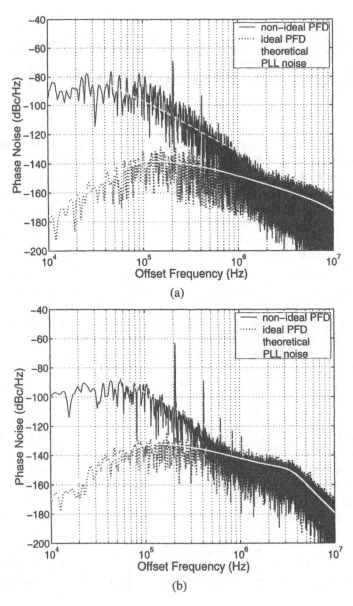

Figure B.1: *Matlab simulation results: The $\Delta\Sigma$ noise at the output of the PLL for (a) the MASH modulator and (b) the single-loop, multi-bit modulator. The results are plotted for an ideal PFD (dotted), which closely corresponds to the theoretical results (solid light grey) and for a non-linear PFD (solid). The results are compared to the simulated PLL phase noise without $\Delta\Sigma$ (the dash-dotted line).*

are the spurious tones, appearing at 200 kHz and harmonics. The fractional part of the division is so close to an integer that the $\Delta\Sigma$ modulators have a hard time randomizing their outputs. This clearly shows up when the discrete-time autocorrelation estimate (Eq. (6.38)) of the outputs is calculated; Large peaks can be distinguished in the autocorrelation estimate for divisions close to integer multiples of f_{ref}, even when dithering is applied. To further investigate the division by 67.9923, an analysis sweep is performed over different degrees of non-linearity. The results are summarized in Fig. B.2. The results for the in-band noise A_n and the phase noise at 600 kHz offset are similar to the results of the analysis in Section 6.7, as could be expected. The results for the spurious tones however are much more pronounced. Even for very small dead zones, the spurious tones are higher than -80 dBc. The effect of gain mismatch is again less severe.

In Fig. B.3, the simulated output phase noise of the synthesizer for a division by 67.577 is presented. This time only noise leakage is observed and spurious tones are well below -80 dBc. For intermediate division values the $\Delta\Sigma$ modulators are able to sufficiently randomize the spurious content of their outputs. Again, an analysis sweep is performed for different degrees of dead zone and gain mismatch. The results for in-band noise and phase noise at 600 kHz, as shown in Fig. B.4, are again similar to that of Section 6.7.

The following conclusions can be drawn:

- The value of the fractional division number has no direct influence on the noise leakage due to non-linearities. Only the change in VCO gain can have a secondary influence, resulting in higher phase noise for lower division numbers.

- For fractional division by numbers in between integer division moduli, no spurious tones are present in the output spectrum of the synthesizer. The $\Delta\Sigma$ modulators are capable of converting any spurious energy into "white" noise.

- The contrary is true for fractional division close to integer division moduli. The $\Delta\Sigma$ modulator is unable to sufficiently randomize its output, as can be expected from Fig. A.2 (b); If the case of the first order modulator with DC-inputs is extrapolated to higher order modulators in terms of correlation in the output, one can notice a high spurious content close to 0 and 1 and close to 0.5. Indeed, with these inputs to the $\Delta\Sigma$ modulator, spurious tones show up at the synthesizer output, although dithering is applied. As stated before, this operating region of the synthesizer should be avoided or much care has to be taken to make a highly linear $\Delta\theta \rightarrow I_{qp}$ conversion, especially in the case of a dead zone.

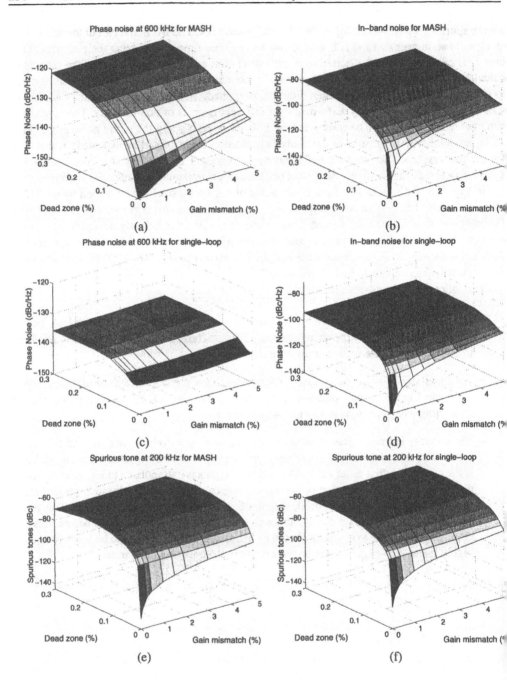

Figure B.2: *Simulation results: The simulated output phase noise of the MASH $\Delta\Sigma$ modulator (a) at 600 kHz and (b) the in-band phase noise versus gain mismatch and dead zone. The simulated output phase noise of the single-loop $\Delta\Sigma$ modulator (c) at 600 kHz and (d) the in-band phase noise. The simulated spurious tones at 200 kHz for the MASH (e) and the single-loop $\Delta\Sigma$ modulator (f) versus gain mismatch and dead zone for a division by 67.9923*

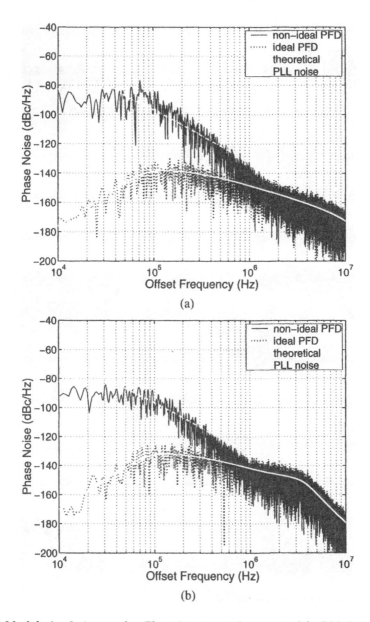

Figure B.3: *Matlab simulation results: The $\Delta\Sigma$ noise at the output of the PLL for (a) the MASH modulator and (b) the single-loop, multi-bit modulator. The results are plotted for an ideal PFD (dotted), which closely corresponds to the theoretical results (solid light grey) and for a non-linear PFD (solid). The results are compared to the simulated PLL phase noise without $\Delta\Sigma$ (the dash-dotted line).*

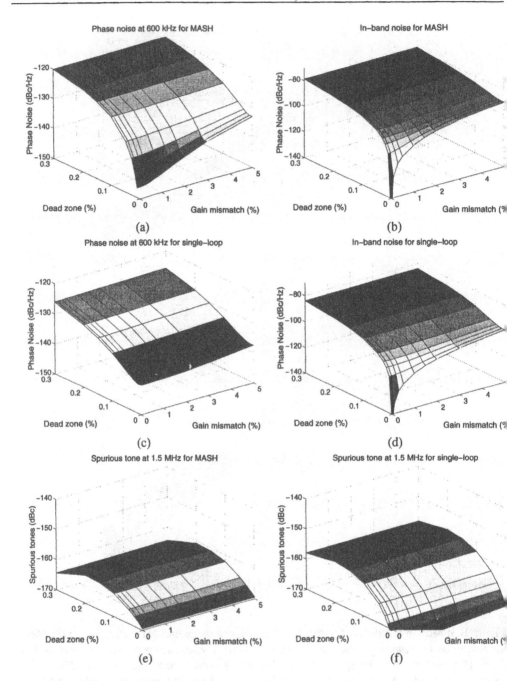

Figure B.4: *Simulation results: The simulated output phase noise of the MASH $\Delta\Sigma$ modulator (a) at 600 kHz and (b) the in-band phase noise versus gain mismatch and dead zone. The simulated output phase noise of the single-loop $\Delta\Sigma$ modulator (c) at 600 kHz and (d) the in-band phase noise. The simulated spurious tones at 1.5 MHz for the MASH (e) and the single-loop $\Delta\Sigma$ modulator (f) versus gain mismatch and dead zone for a division by 67.577*

Index

Bibliography

[Abid83]   A. A. Abidi and R. G. Meyer, "Noise in Relaxation Oscillators", *IEEE Journal of Solid-State Circuits*, vol. 18, no. 6, pp. 794–802, Dec. 1983.

[Abid94]   A. Abidi, "Radio-frequency integrated circuits for portable communications", in *Proceedings Custom Integrated Circuits Conference*, May 1994, pp. 151–158.

[Aero01]   Silicon Laboratories, *Datasheet of the Aero GSM Transceiver Chipset*, http://www.silabs.com/products/areo.asp, 2001.

[Ahol00]   R. Ahola and K. Halonen, "A 2GHz $\Delta\Sigma$ fractional-N Frequency Synthesizer in 0.35 μm CMOS", in *Proceedings European Solid-State Circuits Conference*, Sept. 2000, pp. 472–475.

[Ain00]   H. Ainspan and J.-. Plouchart, "A Comparison of MOS Varactors in Fully-Integrated CMOS LC VCOs at 5 and 7 GHz", in *Proceedings European Solid-State Circuits Conference*, Sept. 2000, pp. 448–451.

[Ajj01]   A. Ajjikuttira, C. Leung, E.-S. Khoo, and M. Choke et. al, "A Fully-Integrated CMOS RFIC for Bluetooth Applications", in *ISSCC, Digest of Technical Papers*, Feb. 2001, pp. 198–199.

[And00]   P. Andreani and S. Matisson, "On the Use of MOS Varactors in RF VCO's", *IEEE Journal of Solid-State Circuits*, vol. 35, no. 6, pp. 905–910, June 2000.

[And01]   P. Andreani and H. Sjoland, "A 2.2GHz CMOS VCO with inductive degeneration noise suppression", in *Proceedings Custom Integrated Circuits Conference*, May 2001, pp. 197–200.

[Arm15]   E. H. Armstrong, "Some Recent Developments in the Audion Receiver", in *IRE Proceedings*, v.3, 1915, pp. 215–247.

[Bark35]   H. Barkhausen, *Lehrbuch der Elektronen-Rohre*, chapter 3.Band, Ruckkopplung, Verlag S. Hirzel, Leipzig, 1935.

[Bel32]   H. de Bellescize, "La réception synchrone", *L'onde électrique*, vol. 11, pp. 225–240, May 1932.

[Best97]   R. E. Best, *Phase-Locked Loops, Design, Simulation & Applications*, McGraw-Hill, 1997.

[Boon89]   C. Boon, *Design of high-performance, negative-feedback oscillators*, Ph.D. Thesis, Technische Universiteit Eindhoven, 1989.

[Borr98]   M. Borremans, B. De Muer, and M. Steyaert, "Phase Noise Up-Conversion Reduction for Integrated CMOS VCOs", *Electronics Letters*, vol. 36, no. 10, pp. 857–858, May 1998.

[Borr00]   M. Borremans, B. De Muer, and M. Steyaert, "The Optimization of GHz Integrated CMOS Quadrature VCOs based on a poly-phase filter loaded differential oscillator", in *Proceedings IEEE International Symposium on Circuits and Systems*, Geneva, May 2000, pp. 729 –732, vol. 2.

[Borr02]   M. Borremans and M. Steyaert, *High Frequency CMOS transmitter integration for wireless and cable applications*, Ph.D. Thesis, K. U. Leuven, 2002.

[Bren02]   G. Brenna, D. Tschopp, D. Pfaff, and Q. Huang, "Circuits A 2GHz Direct-Conversion WCDMA Modulator in 0.25μm CMOS", in *ISSCC, Digest of Technical Papers*, Feb. 2002, pp. 244–245.

[Cand81]   J. C. Candy and O. J. Benjamin, "The structure of quantization noise from Sigma-Delta modulation", *IEEE Transactions on Communications*, vol. 29, pp. 1316–1323, Sept. 1981.

[Chan96]   B. Chang, J. Prak, and W. Kim, "A 1.2 GHz CMOS Dual-Modulus Prescaler Using New Dynamic D-Type Flip-Flops", *IEEE Journal of Solid-State Circuits*, vol. 31, no. 5, pp. 749–752, May 1996.

[Cho99]   T. Cho et al., "A Single-Chip CMOS Direct-Conversion Transceiver for 900MHz Spread Spectrum Digital Cordless Phones", in *ISSCC, Digest of Technical Papers*, Feb. 1999.

[Cong88]   H. Cong, J. Andrews, D. M. Boulin, S.-C. Fang, S. J. Hillenius, and J. A. Michejda, "A 2 GHz CMOS Dual-Modulus Prescaler IC", in *ISSCC, Digest of Technical Papers*, Feb. 1988, pp. 138–139.

[Cox76]   R. Cox, *Patent No. 3,976,945*, Washington DC: US Patent Office, 1976.

[Cran95a]   J. Craninckx and M. S. J. Steyaert, "A 1.8 GHz CMOS Low-Phase-Noise Voltage-Controlled Oscillator with prescaler", *IEEE Journal of Solid-State Circuits*, vol. 30, no. 12, pp. 1474–1482, Dec. 1995.

[Cran95b]   J. Craninckx and M. Steyaert, "Low-noise voltage controlled oscillators using enhanced LC-tanks", *IEEE Transactions on Circuits and Systems II*, vol. 42, no. 12, pp. 743–751, Dec. 1995.

[Cran97a] J. Craninckx and M. Steyaert, "A fully integrated spiral-LC CMOS VCO set with prescaler for GSM and DCS-1800 systems", in *Proceedings Custom Integrated Circuits Conference*, May 1997, pp. 403–406.

[Cran97b] J. Craninckx and M. S. J. Steyaert, "A 1.8 GHz low-phase-noise CMOS VCO using optimized hollow inductors", *IEEE Journal of Solid-State Circuits*, vol. 32, no. 4, pp. 736–744, May 1997.

[Cran98] J. Craninckx and M. Steyaert, *Wireless CMOS Frequency Synthesizer Design*, Kluwer Academic Publishers, 1998.

[Crol95] J. Crols and M. Steyaert, "A single-chip 900 MHz CMOS receiver front-end with a high performance low-IF topology", *IEEE Journal of Solid-State Circuits*, vol. 30, no. 12, pp. 1483–1492, Dec. 1995.

[Crol96] J. Crols, P. Kinget, J. Craninckx, and M. Steyaert, "An analytical model for planar inductors on lowly doped silicon substrates for high frequency analog design up to 3 GHz", in *Digest of Technical Papers, Symposium on VLSI Circuits*, June 1996, pp. 28–29.

[CYC00] C. De Ranter, B. De Muer, G. Van der Plas, P. Vancorenland, M. Steyaert, G. Gielen, and W. Sansen, "CYCLONE: Automated Design and Layout of RF LC-Oscillators", in *Proceedings Design Automation Conference*, June 2000, pp. 11–14.

[Dar01] H. Darabi, S. Khorram, E. Chien, and M. Pan et. al, "A 2.4GHz CMOS Transceiver for Bluetooth", in *ISSCC, Digest of Technical Papers*, Feb. 2001, pp. 198–199.

[Dec99] A. Dec and K. Suyama, "A 1.9 GHz Micromachined-based Low-Phase-Noise CMOS VCO", in *ISSCC, Digest of Technical Papers*, bondwire, Feb. 1999, pp. 80–81.

[DeCo01] W. De Cock and M. Steyaert, "A CMOS 10GHz Voltage Controlled LC-Oscillator with Integrated high-Q Inductor", in *Proceedings European Solid-State Circuits Conference*, Villach, Austria, Sept. 2001, pp. 496–499.

[Demi98] A. Demir, A. Mehrotra, and J. Roychowdhury, "Phase Noise and Timing Jitter in Oscillators", in *Proceedings Custom Integrated Circuits Conference*, Santa Clara, USA, May 1998, pp. 45–48.

[DeMu98] B. De Muer and M. Steyaert, "A Single-ended 1.5 GHz 8/9 Dual-modulus Prescaler in 0.7μm CMOS with Low-Phase-Noise and High Input Sensitivity", in *Proceedings European Solid-State Circuits Conference*, The Hague, Sept. 1998, pp. 256–259.

[DeMu99a] B. De Muer, C. D. Ranter, and M. Steyaert, "A fully integrated 2GHz LC-VCO with phase noise of -125dBc/Hz at 600kHz", in *Proceedings European Solid-State Circuits Conference*, Duisburg, Sept. 1999, pp. 206–209.

[DeMu99b] B. De Muer, C. De Ranter, J. Crols, and M. Steyaert, "A Simulator-Optimizer for the design of very low-phase-noise CMOS LC-oscillators", in *ICECS*, Pahpos, Sept. 1999, pp. 1557–1560, Vol. III.

[DeMu00a] B. De Muer, M. Borremans, N. Itoh, and M. Steyaert, "A 1.8 GHz highly-tunable, low-phase-noise CMOS VCO", in *Proceedings Custom Integrated Circuits Conference*, Orlando, May 2000, pp. 585–588.

[DeMu00b] B. De Muer, M. Borremans, and M. Steyaert, "A 2GHz low-phase-noise integrated LC-VCO set with flicker noise upconversion minimization", *IEEE Journal of Solid-State Circuits*, vol. 35, no. 7, pp. 1034–1038, July 2000.

[DeMu00c] B. De Muer and M. Steyaert, "A 12GHz /128 Frequency Divider in 0.25µm CMOS", in *Proceedings European Solid-State Circuits Conference*, Stockholm, Sept. 2000, pp. 220–223.

[DeMu00d] B. D. Muer and M. Steyaert, *Fully Integrated CMOS Frequency Synthesizers for Wireless Communications*, pp. 287–323, Analog Circuit Design, W. Sansen, J. H. Huijsing, R. J. van de Plassche (eds.) Kluwer Academic Publishers, 2000.

[DeMu02] B. De Muer and M. Steyaert, "A CMOS Monolithic Delta-Sigma-controlled Fractional-N Frequency Synthesizer for DCS-1800", *IEEE Journal of Solid-State Circuits*, vol. 37, no. 7, pp. 835–844, July 2002.

[DeRa01] C. De Ranter and M. Steyaert, "A 0.25m CMOS 17GHz VCO", in *ISSCC, Digest of Technical Papers*, Feb. 2001, pp. 469–470.

[DeSm98] B. De Smedt and G. Gielen, "On the Evaluation of Phase Noise in Frequency Dividers", in *Proceedings European Solid-State Circuits Conference*, Sept. 1998, pp. 260–263.

[EDG99] NOKIA, *Enhanced Data Rates for GSM Evolution EDGE*, White paper, 1999.

[Egan81] W. Egan, *Frequency Synthesis by Phase Lock*, J. Wiley & Sons, New York, USA, 1981.

[Egan90] W. Egan, "Modeling phase noise in frequency dividers", *IEEE Transactions on Ultrasonics, Ferroelectrics and Frequency Control*, vol. 37, no. 4, pp. 307–315, July 1990.

[Enz95] C. Enz, F. Krummenacher, and E. A. Vittoz, "An analytical MOS transistor model valid in all regions of operation and dedicated to low-voltage and low-current applications", *Analog Integrated Circuits and Signal Processing Journal on Low-Voltage and Low-Power Design*, vol. 8, no. 1, pp. 83–114, July 1995.

[ETS00]   ETSI EN 300 908 (GSM 05.02 version 8.5.1 Release 1999), *Digital cellular communication system (Phase 2+); Multiplexing and multiple access on the radio path*, European Telecommunications Standards Institute, 2000.

[ETSI00]   ETSI EN 300 190 (GSM 05.05 version 8.5.1 Release 1999), *Digital cellular communication system (Phase 2+); Radio transmission and reception*, European Telecommunications Standards Institute, 2000.

[Eur93]   EuropTest, *Phase Noise Theory and Measurement*, chapter 93-0001, Application Note Europtest, 1993.

[Eynd93]   F. Op 't Eynde and W. Sansen, *Analog Interfaces for Digital Signal Processing Systems*, chapter 3, pp. 83–123, Kluwer Academic Publishers, 1993.

[Eynd01]   F. Op 't Eynde, Jean-Jacques Schmit, and Vincent Charlier et. al, "A Fully-Integrated Single-Chip SoC for Bluetooth", in *ISSCC, Digest of Technical Papers*, Feb. 2001, pp. 196–197.

[Fast96]   M. Kamon and L. M. Silveira et al., *FastHenry USER'S GUIDE: version 3.0*, Massachusetts Institute of Technology, 1996.

[Fili98]   N. M. Filiol, T. A. D. Riley, C. Plett, and M. A. Copeland, "An Agile ISM Band Frequency Synthesizer with Built-in GMSK Data Modulation", *IEEE Journal of Solid-State Circuits*, vol. 33, no. 7, pp. 998–1008, July 1998.

[Fong02]   N. Fong, J. Plouchart, and N. Z. et al., "A Low-Voltage Multi-GHz VCO with 58% tuning range in SOI CMOS", in *Proceedings Custom Integrated Circuits Conference*, May 2002, pp. 423–426.

[Fran93]   J. Franca and Y. Tsividis, *Design of Analog-Digital VLSI Circuits for Telecommunications and Signal Processing*, chapter 17, Prentice Hall, 1993.

[Fren01]   L. Frenzel, "Transceiver Chip Set Wrings Out GSM Phone Costs", *Electronic Design*, vol. 49, no. 5, Mar. 2001.

[Gard79]   F. Gardner, *Phaselock Techniques*, J. Wiley & Sons, New York, USA, 1979.

[Gee00]   Y. Geerts, M. Steyaert, and W. Sansen, "A High-Performance Multi-Bit CMOS $\Delta\Sigma$ Converter", *IEEE Journal of Solid-State Circuits*, vol. 35, no. 12, pp. 1829–1840, Dec. 2000.

[Gier99]   S. Gierkink, E. Klumperink, A. van der Wel, G. Hoogzaad, E. van Tuijl, and B. Nauta, "Intrinsic 1/f device noise reduction and its effect on phase noise in CMOS ring oscillators", *IEEE Journal of Solid-State Circuits*, vol. 34, no. 7, pp. 1022–1025, July 1999.

[Gold95]   B. Goldberg, *Digital techniques in frequency synthesis*, McGraw-Hill, 1995.

[Gray89]   R. M. Gray, "Spectral analysis of quantization noise in a single loop sigma-delta modulator with dc input", *IEEE Transactions on Communications*, vol. 37, pp. 588–599, June 1989.

[Gree74]   H. M. Greenhouse, "Design of planar rectangular microelectronic inductors", *IEEE Transactions on Parts, Hybrids and Packaging*, vol. 10, no. 2, pp. 101–109, June 1974.

[Haji98]   A. Hajimiri and T. Lee, "A general theory of phase noise in electrical oscillators", *IEEE Journal of Solid-State Circuits*, vol. 33, no. 2, pp. 179–194, Feb. 1998.

[Haji99]   A. Hajimiri, S. Limotyrakis, and T. H. Lee, "Jitter and Phase Noise in Ring Oscillators", *IEEE Journal of Solid-State Circuits*, vol. 34, no. 6, pp. 790–804, June 1999.

[Heg01]   E. Hegazi, H. Sjoland, and A. Abidi, "A filtering technique to lower oscillator phase noise", in *ISSCC, Digest of Technical Papers*, Feb. 2001, pp. 364–365.

[Herz00]   F. Herzel, M. Pierschel, P. Weger, and M. Tiebout, "Phase Noise in a Differential CMOS Voltage-Controlled Oscillator for RF Applications", *IEEE Transactions on Circuits and Systems II*, vol. 47, no. 1, pp. 11–15, Jan. 2000.

[Hill92]   A. Hill and J. Surber, "The PLL dead zone and how to avoid it", *RF Design*, pp. 131–134, Mar. 1992.

[Hosh01]   K. Hoshino, E. Hegazi, J. Rail, and A. Abidi, "A 1.5V, 1.7 mA 700 MHz CMOS LC oscillator with no upconverted flicker noise", in *Proceedings European Solid-State Circuits Conference*, Villach, Austria, Sept. 2001, pp. 352–355.

[HP93]   –, *HP80000 Data Generator System: Installation Guide*, Hewlett-Packard (now Agilent), 1993.

[Hua00]   Q. Huang, "Phase noise to carrier ratio in LC oscillators", *IEEE Transactions on Circuits and Systems I*, vol. 47, no. 7, pp. 965–980, July 2000.

[Huan96]   Q. Huang and R. Rogenmoser, "Speed Optimization of Edge-Triggered CMOS Circuits for Gigahertz Single-Phase Clocks", *IEEE Journal of Solid-State Circuits*, vol. 31, no. 3, pp. 456–465, Mar. 1996.

[Ing89]   A. Ingber, "Very Fast Simulated Re-Annealing", *Mathematical Computer Modelling*, vol. 12, no. 8, pp. 976–973, 1989.

[ITRS99]   *International Technology Roadmap for Semiconductors*, Technology working groups, Austin, TX, 1999.

[Jans02]   J. Janssens and M. Steyaert, *CMOS Cellular Receiver Front-Ends from Specification to Realization*, Kluwer Academic Publishers, 2002.

[John97]   D. A. Johns and K. Martin, *Analog Integrated Circuit design*, J. Wiley & Sons, 1997.

[Kado93]   Y. Kado, M. Suzuki, K. Koike, Y. Omura, and K. Izumi, "A 1 GHz/0.9 mW CMOS/SIMOX Divide-by-128/129 Dual-Modulus Prescaler Using a Divide-by-2/3 Synchronous Counter", *IEEE Journal of Solid-State Circuits*, vol. 28, no. 4, pp. 513–517, Apr. 1993.

[Kara96]   A. Karanicolas, "A 2.7V 900MHz CMOS LNA and Mixer", *IEEE Journal of Solid-State Circuits*, vol. 31, no. 12, pp. 1939–1944, Dec. 1996.

[Kim00]   J. Kim and B. Kim, "A Low-Phase-Noise CMOS LC Oscillator with a Ring Structure", in *ISSCC, Digest of Technical Papers*, CMOS ringLC, Feb. 2000, pp. 518–519.

[Kin98a]   P. Kinget, "A fully integrated 2.7V 0.35μm CMOS VCO for 5GHz Wireless Applications", in *ISSCC, Digest of Technical Papers*, San Fransisco, USA, Feb. 1998, pp. 226–227.

[Kin98b]   P. Kinget, "A 2.4 GHz CMOS VCO with MCM-inductor", in *Proceedings European Solid-State Circuits Conference*, Sept. 1998, pp. 364–367.

[Kin99]   P. Kinget, *Integrated GHz Voltage Controlled Oscillators*, chapter Analog Circuit Design, W. Sansen, J. H. Huijsing, R. J. van de Plassche (eds.) Kluwer Academic Publishers, pp. 353–381, New York : Kluwer Academic Publishers, 1999.

[King75]   C. A. Kingsford-Smith, *Patent No. 3,928,813*, Washington DC: US Patent Office, 1975.

[Kou93]   I. Koullias, J. Havens, I. Post, and P. E. Bronner, "A 900 MHz Transceiver Chip Set for Dual-Mode Cellular Radio Mobile Terminals", in *ISSCC, Digest of Technical Papers*, Feb. 1993, pp. 140–141.

[Kral98]   A. Kral, F. Behbahani, and A. A. Abidi, "RF-CMOS Oscillators with Switched Tuning", in *Proceedings Custom Integrated Circuits Conference*, May 1998, pp. 26.1.1 – 26.1.4.

[Kuc01a]   J. Kucera, "Wideband BiCMOS VCO for GSM/UMTS Direct Conversion Receivers", in *ISSCC, Digest of Technical Papers*, bondwire, Feb. 2001, pp. 374–375.

[Kuc01b]   J. Kucera and B. Klepser, "3.6 GHz VCOs for multi-band GSM transceivers", in *Proceedings European Solid-State Circuits Conference*, Villach, Austria, Sept. 2001, pp. 340–343.

[Kuhn95]   W. Kuhn, A. Elshabini-Riad, and F. Stephenson, "Centre-tapped spiral inductors for monolithic bandpass filters", *Electronics Letters*, vol. 31, no. 8, pp. 625–626, Apr. 1995.

[Kur97]   M. Kurisu, M. Nishikawa, and A. e. a. H, "An 11.8 GHz 31 mW CMOS Frequency Divider", in *Digest of Technical Papers, Symposium on VLSI Circuits*, 1997, pp. 73–74.

[Lars96]   P. Larsson, "High-speed Architecture for a Dual-Modulus Prescaler and a Programmable Frequency Divider", *IEEE Journal of Solid-State Circuits*, vol. 31, no. 5, pp. 744–748, May 1996.

[Lee98a]   J. Lee, A. Kral, and A. Abidi, "Design of Spiral Inductors on Silicon Substrates with a Fast Simulator", in *Proceedings European Solid-State Circuits Conference*, Sept. 1998, pp. 328–331.

[Lee98b]   T. H. Lee, *The Design of CMOS Radio-Frequency Integrated Circuits*, chapter 1, Cambridge University Press, 1998.

[Lee98c]   T. Lee, *The Design of CMOS Radio-Frequency Integrated Circuits*, chapter 2, Cambridge University Press, 1998.

[Lee01]   S. Lee, M. Yoh, J. Lee, and I. Ryu, "A 17 mW, 2.5 GHz Fractional-N Frequency Synthesizer for CDMA-2000", in *Proceedings European Solid-State Circuits Conference*, Villach, Sept. 2001, pp. 40–43.

[Leen02]   D. Leenaerts, C. Vaucher, and H.J. Bergveld et al., "A 15mW Fully Integrated I/Q Synthesizer for Bluetooth in 0.18μm CMOS", in *Proceedings European Solid-State Circuits Conference*, Firenze, Italy, Sept. 2002, pp. 93–96.

[Lees66]   D. Leeson, "A Simple Model of Feedback Oscillator Noise Spectrum", *IEEE Proceedings*, vol. 54, pp. 329–330, Feb. 1966.

[LeGi01]   D. Leenaerts, G. Gielen, and R. Rutenbar, "CAD solutions and outstanding challenges for mixed-signal and RF IC design", in *Proceedings IEEE/ACM International Conference on Computer Aided Design*, Nov. 2001, pp. 270–277.

[Lev02]   S. Levantino, C. Samori, A. Bonfanti, S. Gierkink, A. Laicata, and V. Boccuzzi, "Frequency dependence on bias current in 5-GHz CMOS VCOs: Impact on tuning range and flicker noise upconversion", *IEEE Journal of Solid-State Circuits*, vol. 37, no. 8, pp. 1003–1011, Aug. 2002.

[Lin00]   J. Lin, "An Integrated Low-Phase-Noise Voltage Controlled Oscillator for Base Station Applications", in *ISSCC, Digest of Technical Papers*, bipolar, Feb. 2000, pp. 519–520.

[Liu00]   T.-P. Liu, E. Westerwick, N. Rohani1, and R.-H. Yan, "5GHz CMOS Radio Transceiver Front-End Chipset", in *ISSCC, Digest of Technical Papers*, Feb. 2000, pp. 320–321.

[Lo02]   C.-W. Lo and H. C. Luong, "A 1.5 V 900 MHz Monolithic CMOS Fast-Switching Frequency Synthesizer for Wireless Applications", *IEEE Journal of Solid-State Circuits*, vol. 37, no. 4, pp. 459–470, Apr. 2002.

[Mag93]   E. Freeman, *MagNet 5 user guide – Using the MagNet version 5 package from Infolytica*, Infolytica, 1993.

[Man02]   D. Manstretta, R. Castello, F. Gatta, P. Rossi, and F. Svelto, "Circuits A 0.18μm CMOS Direct-Conversion Receiver Front-End for UMTS", in *ISSCC, Digest of Technical Papers*, Feb. 2002, pp. 240–241.

[Mars95]   C. Marshall et al., "2.7 V GSM Transceiver ICs with On-Chip Filtering", in *ISSCC, Digest of Technical Papers*, Feb. 1995, pp. 148–149.

[Mat97]   The Mathworks, *Matlab user's guide, version 5*, Prentice Hall, 1997.

[Mat02]   A. Matsuzawa, "RF-SoC, Expectations and Required Conditions", *IEEE Transactions on Microwave Theory and Techniques*, vol. 50, no. 1, pp. 26, Jan. 2002.

[McCl92]   M. McClure, "Residual phase noise of digital frequency dividers", *Microwave Journal*, pp. 124–130, Mar. 1992.

[Miju94]   D. Mijuskovic and M. Bayer, "Cell based fully integrated CMOS frequency synthesizers", *IEEE Journal of Solid-State Circuits*, vol. 29, no. 3, pp. 271–279, Mar. 1994.

[Mill90]   B. Miller and B. Conley, "A Multiple Modulator Fractional Divider", in *44th Annual Symposium on Frequency Control*, 1990, pp. 559–568.

[Mill91]   B. Miller and R. Conley, "A Multiple Modulator Fractional Divider", *IEEE Transactions on Instrumentations and Measurements*, vol. 40, no. 3, pp. 578–583, June 1991.

[Moh99]   S. S. Mohan, M. Hershenson, S. Boyd, and T. Lee, "Simple Accurate Expressions for Planar Spiral Inductors", *IEEE Journal of Solid-State Circuits*, vol. 34, no. 10, pp. 1419–1424, Oct. 1999.

[MOSC]   P. Vancorenland, *MOSCAL v1.6 tutorial*, http://www.esat.kuleuven.ac.be/~vanco/moscal.

[Nik98]   A. M. Niknejad and R. Meyer, "Analysis, Design and Optimization of Spiral Inductors and Transformers for Si RF IC's", *IEEE Journal of Solid-State Circuits*, vol. 33, no. 10, pp. 1470–1481, Oct. 1998.

[Nik99]   A. M. Niknejad, R. G. Meyer, and J. L. Tham, "Fully-Integrated Low Phase Noise Bipolar Differential VCOs at 2.9 and 4.4 GHz", in *Proceedings European Solid-State Circuits Conference*, Sept. 1999, pp. 198–201.

[Nors97]   S. R. Norsworthy, R. Schreier, and G. C. Themes, *Delta-Sigma Data Converters: Theory, Design and Simulation*, IEEE Press, 1997.

[Park99]   C.-H. Park and B. Kim, "A low-nosie, 900 MHz VCO in 0.6 μm CMOS", *IEEE Journal of Solid-State Circuits*, vol. 34, no. 5, pp. 586–591, May 1999.

[Ped99]   E. Pederson, "Performance Evaluation of CMOS Varactors for Wireless RF Applications", in *Proceedings of the 17th Nordic Conference*, Oslo, Norway, Nov. 1999, pp. 73–78.

[Perr97]   M. H. Perrott, *Techniques for High Data Rate Modulation and Low Power Operation of Fractional-N Frequency Synthesizers*, Ph.D. Thesis, Massachusetts Institute of Technology (MIT), 1997.

[Pfaf99]   D. Pfaff and Q. Huang, "A Quarter-Micron CMOS, 1 GHz VCO/Prescaler Set for very low power applications", in *Proceedings Custom Integrated Circuits Conference*, external L, May 1999, pp. 649–652.

[Pfaf02]   D. Pfaff and Q. Huang, "An 18 mW 1800MHz quadrature demodulator in 0.18μm CMOS", in *ISSCC, Digest of Technical Papers*, San Fransisco, USA, Feb. 2002, pp. 242–243.

[Ping99]   T. Ping, "A 6.5 GHz Monolithic CMOS Voltage-Controlled Oscillator", in *ISSCC, Digest of Technical Papers*, Feb. 1999, pp. 404–405.

[Porr00]   A.-S. Porret, T. Melly, C. Enz, and E. Vittoz, "Design of High-Q Varactors for Low-Power Wireless Applications Using a Standard CMOS Process", *IEEE Journal of Solid-State Circuits*, vol. 35, no. 3, pp. 337–345, Mar. 2000.

[Post98]   J. Post, I. Linscott, and M. Oslick, "Waveform symmetry properties and phase noise in oscillators", *Electronics Letters*, vol. 34, no. 16, pp. 1547–1548, Aug. 1998.

[Rab02]   J. Rabaey, "ESSCIRC Workshop: Electrical Issues in SoC/SoP Design", in *Proceedings European Solid-State Circuits Conference*, Sept. 2002.

[Raz95]   B. Razavi, K. F. Lee, and R. H. Yan, "Design of High-Speed, Low-Power Frequency Dividers and Phase-Locked Loops in Deep Submicron CMOS", *IEEE Journal of Solid-State Circuits*, vol. 30, no. 2, pp. 101–109, Feb. 1995.

[Raz96]   B. Razavi, "A study of phase noise in CMOS oscillators", *IEEE Journal of Solid-State Circuits*, vol. 31, no. 3, pp. 331–343, Mar. 1996.

[Raz97a]   B. Razavi, "A 1.8 GHz CMOS voltage controlled oscillator", in *ISSCC, Digest of Technical Papers*, Feb. 1997, pp. 388–389.

[Raz97b]   B. Razavi, "Challenges in the design of frequency synthesizers for wireless applications", in *Proceedings Custom Integrated Circuits Conference*, Santa Clara, USA, May 1997, pp. 395–402.

Rhee00a] W. Rhee, B. Bisanti, and A. Ali, "A 18 mW, 2.5 GHz/900 MHz BiCMOS Dual Frequency Synthesizer with < 10Hz RF Carrier Resolution", in *Proceedings European Solid-State Circuits Conference*, Sept. 2000, pp. 224–227.

Rhee00b] W. Rhee, B.-S. Song, and A. Ali, "A 1.1-GHz CMOS Fractional-N Frequency Synthesizer with a 3-b Third-Order $\Delta\Sigma$ Modulator ", *IEEE Journal of Solid-State Circuits*, vol. 35, no. 10, pp. 1453–1460, Oct. 2000.

Rhod83] U. L. Rhode, *Digital PLL Frequency Synthesizers, Theory and Design*, Prentice-Hall Inc., Englewood Cliffs, USA, 1983.

Ril93] T. Riley, M. Copeland, and T. Kwasniewski, "Delta-sigma modulation in fractional-N frequency synthesis", *IEEE Journal of Solid-State Circuits*, vol. 28, no. 5, pp. 553–559, May 1993.

Rof96a] A. Rofourgan, J. Rael, M. Rofourgan, and A. Abidi, "A 900 MHz CMOS LC-oscillator with quadrature outputs", in *ISSCC, Digest of Technical Papers*, Feb. 1996, pp. 392–393.

Rof96b] A. Rofougaran, J. Chang, M. Rofougaran, and A. Abidi, "A 1GHz CMOS RF front-end IC for a Direct Conversion Wireless Receiver", *IEEE Journal of Solid-State Circuits*, vol. 31, no. 7, pp. 880–889, July 1996.

Rof98a] A. Rofougaran et al., "A Single-Chip 900 MHz Spread-Spectrum Wireless Transceiver in $1\mu m$ CMOS – Part I: Architecture and Transmitter Design", *IEEE Journal of Solid-State Circuits*, vol. 33, no. 4, pp. 515–534, Apr. 1998.

Rof98b] A. Rofougaran et al., "A Single-Chip 900 MHz Spread-Spectrum Wireless Transceiver in $1\mu m$ CMOS – Part II: Receiver Design", *IEEE Journal of Solid-State Circuits*, vol. 33, no. 4, pp. 535–547, Apr. 1998.

Rog94] R. Rogenmoser, Q. Huang, and F. Piazza, "1.57 GHz asynchronous and 1.4 GHz dual-modulus $1.2\mu m$ CMOS prescalers", in *Proceedings Custom Integrated Circuits Conference*, May 1994, pp. 387–390.

Sab96] Analogy, *The Saber User Manual*, 1996.

Samo98] C. Samori, F. Villa, and F. Zappa, "Spectrum Folding and Phase Noise in LC Tuned Oscillators", *IEEE Transactions on Circuits and Systems II*, vol. 45, no. 7, pp. 781–790, July 1998.

San84] J. T. Santos and R. G. Meyer, "A one-pin crystal oscillator for VLSI circuits", *IEEE Journal of Solid-State Circuits*, vol. 19, no. 2, pp. 228–236, Apr. 1984.

Sans94] K. R. Laker and W. M. C. Sansen, *Design of Analog Integrated Circuits and Systems*, chapter 2, McGraw-Hill, 1994.

[Sen89]   R. Senani and B. A. Kumar, "Linearly tunable Wien bridge oscillator realized with operation transconductance amplifiers", *Electronics Letters*, vol. 25, no. 1, pp. 19–21, Jan. 1989.

[Sev91]   J. Sevenhans et al., "An integrated Si bipolar RF transceiver for a zero IF 900MHz GSM digital mobile radio frontend of a hand portable phone", in *Proceedings Custom Integrated Circuits Conference*, May 1991.

[Sev94]   J. Sevenhans et al., "An Analog Radio Front-End Chip Set for a 1.9GHz Mobile Radio Telephone Application", in *ISSCC, Digest of Technical Papers*, Feb. 1994, pp. 44–45.

[Sil02]   Silicon Strategies, *Monthly outlook: where IC markets are headed*, http://www.siliconstrategies.com, 2002.

[Snee90]   J. Sneep and C. Verhoeven, "A new low-noise 100 MHz balanced relaxation oscillator", *IEEE Journal of Solid-State Circuits*, vol. 25, no. 3, pp. 692–698, June 1990.

[Soy89]   M. Soyeur and R. Meyer, "High frequency phase-locked loops in monolithic bipolar technology", *IEEE Journal of Solid-State Circuits*, vol. 24, no. 3, pp. 787–795, June 1989.

[Soy90]   M. Soyuer and R. G. Meyer, "Frequency limitations of a conventional phase-frequency detector", *IEEE Journal of Solid-State Circuits*, vol. 25, no. 4, pp. 1019–1022, Aug. 1990.

[Stet95]   T. Stetzler, I. Post, J. Havens, and M. Koyama, "A 2.7-4.5V single chip GSM transceiver RF integrated circuit", *IEEE Journal of Solid-State Circuits*, vol. 30, no. 12, pp. 1421–1429, Dec. 1995.

[Stey93]   M. Steyaert and W. Sansen, "Opamp design toward maximum gain-bandwidth", in *Proceedings Workshop on Advances in Analog Circuit Design*, Mar. 1993, pp. 63–85.

[Stey98]   M. Steyaert, M. Borremans, J. Janssens, B. De Muer, and N. Itoh et al., "A Single-Chip CMOS Transceiver for DCS-1800 Wireless Communications", in *ISSCC, Digest of Technical Papers*, San Francisco, Feb. 1998, pp. 48–49.

[Stey00a] M. Steyaert, J. Janssens, B. De Muer, M. Borremans, and N. Itoh, "A 2V CMOS cellular transceiver front-end", in *ISSCC, Digest of Technical Papers*, San Francisco, Feb. 2000, pp. 142–143.

[Stey00b] M. Steyaert, J. Janssens, B. De Muer, M. Borremans, and N. Itoh, "A 2V CMOS cellular transceiver front-end", *IEEE Journal of Solid-State Circuits*, vol. 35, no. 12, pp. 1895–1907, Dec. 2000.

Su02]   D. Su, M. Zargari, and Patrick Yue et al., "A 5GHz CMOS Transceiver for IEEE
            802.11a Wireless LAN", in *ISSCC, Digest of Technical Papers*, Feb. 2002, pp.
            92–93.

Svel00]   F. Svelto, S. Deantoni, and R. Castello, "A 1mA, -120dBc/Hz at 600kHz from 1.9
            GHz fully tuneable LC CMOS VCO", in *Proceedings Custom Integrated Circuits
            Conference*, Orlando, USA, May 2000, pp. 577–580.

Tang99]   J. van der Tang and S. Hahn, "A monolithic 0.4mW SOA LC Voltage Controlled
            Oscillator", in *Proceedings European Solid-State Circuits Conference*, Sept. 1999,
            pp. 150–153.

Tie02]   M. Tiebout, H.-D. Wohlmuth, and W. Simburger, "A 1 V 51GHz Fully-integrated
            VCO in 0.12m CMOS", in *ISSCC, Digest of Technical Papers*, Feb. 2002, pp.
            300–301.

TSB98]   Internal Toshiba Report, *Transceiver simulation using the Advanced Design System
            (ADS) of Agilent Technologies*, 1998.

Vaa01]   P. Vaananen, M. Metsanvirta, and N. Tchamov, "A 4.3 GHz VCO with 2 GHz
            tuning range and low phase noise", *IEEE Journal of Solid-State Circuits*, vol. 36,
            no. 1, pp. 142–146, Jan. 2001.

Vauc01]   C. Vaucher, *Architecture for RF synthesizers*, Ph.D. thesis, University of Twente,
            2001.

Vit88]   E. Vittoz, M. Degrauwe, and S. Bitz, "High-performance crystal oscillator: Theory
            and Application", *IEEE Journal of Solid-State Circuits*, vol. 23, no. 3, pp. 774–783,
            June 1988.

Wak98]   T. Wakimoto and S. Konaka, "A 1.9 GHz Si Bipolar Quadrature VCO wit Fully-
            Integrated LC-Tank", in *Digest of Technical Papers, Symposium on VLSI Circuits*,
            June 1998, pp. 30–31.

Wang90]   Y.-T. Wang and A.A.Abidi, "CMOS active filter design at high frequencies", *IEEE
            Journal of Solid-State Circuits*, vol. 25, no. 6, pp. 1562–1574, Dec. 1990.

Wang98]   H. Wang, "A 9.8 GHz Back-Gate Tuned VCO in 0.35μm CMOS", in *ISSCC,
            Digest of Technical Papers*, Feb. 1998, pp. 406–407.

Wang00]   H. Wang, "A 1.8 V 3mW 16.8 GHz Frequency Divider in 0.25μm CMOS", in
            ISSCC, Digest of Technical Papers, feb 2000, pp. 196–197.

Wang01]   H. Wang, "A 50GHz VCO in 0.25m CMOS", in *ISSCC, Digest of Technical Papers*,
            Feb. 2001, pp. 470–471.

[Wei94]   T. C. Weigandt, B. Kim, and P. R. Gray, "Analysis of timing jitter in CMOS ring os-
            cillators", in *Proceedings IEEE International Symposium on Circuits and Systems*,
            London, GB, May 1994, pp. 27–30.

[Will00]   S. Willingham, M. Perrott, B. Setterberg, A. Grzegorek, and B. McFarland, "An
            integrated 2.5GHz $\Delta\Sigma$ frequency synthesizer with $5\mu s$ settling and 2Mb/s closed
            loop modulation", in *ISSCC, Digest of Technical Papers*, Feb. 2000, pp. 200–201.

[Wong00]   W. Wong, P. Hui, Z. Cheng, K. Shen, J. Lau, P. Chan, and P.-K. Ko, "A Wide
            Tuning Range Gated Varactor", *IEEE Journal of Solid-State Circuits*, vol. 35, no.
            5, pp. 773–779, May 2000.

[Wu00]   H. Wu and A. Hajimiri, "A 10 GHz CMOS distributed voltage controlled oscilla-
            tor", in *Proceedings Custom Integrated Circuits Conference*, May 2000, pp. 581–
            584.

[Yim01]   S.-M. Yim and K. K. O, "Demostration of a switched resonator concept in a dual-
            band monolithic CMOS LC-Tuned VCO", in *Proceedings Custom Integrated Cir-
            cuits Conference*, May 2001, pp. 205–208.

[Yuan89]   J. Yuan and C. Svensson, "High-speed CMOS Circuit Technique", *IEEE Journal
            of Solid-State Circuits*, vol. 24, no. 1, pp. 62–70, Jan. 1989.

[Yuan97]   J. Yuan and C. Svensson, "New single-clock CMOS latches and flipflops with
            improved speed and power savings", *IEEE Journal of Solid-State Circuits*, vol. 32,
            no. 1, pp. 62–67, Jan. 1997.

[Yue98]   C. P. Yue and S. S. Wong, "On-Chip Spiral Inductors with Patterned Ground Shields
            for Si-Based RF IC's", *IEEE Journal of Solid-State Circuits*, vol. 33, no. 5, pp. 743–
            752, May 1998.

[Zan01]   A. Zanchi, A. Bonfanti, S. Levantino, C. Samori, and A. Lacaita, "Automatic Am-
            plitude Control Loop for a 2V, 2.5 GHz LC-tank VCO", in *Proceedings Custom
            Integrated Circuits Conference*, May 2001, pp. 209–211.

Printed in the United States
By Bookmasters